THE GENETIC LOTTERY

The Genetic Lottery

Why DNA Matters for Social Equality

Kathryn Paige Harden

PRINCETON UNIVERSITY PRESS

PRINCETON AND OXFORD

Published by Princeton University Press
41 William Street, Princeton, New Jersey 08540
6 Oxford Street, Woodstock, Oxfordshire OX20 1TR

press.princeton.edu

All Rights Reserved

Library of Congress Cataloging-in-Publication Data

Names: Harden, Kathryn Paige, author.
Title: The genetic lottery : why DNA matters for social equality / Kathryn Paige Harden.
Description: Princeton : Princeton University Press, [2021] | Includes bibliographical references and index.
Identifiers: LCCN 2021012198 (print) | LCCN 2021012199 (ebook) |
    ISBN 9780691190808 (hardback) | ISBN 9780691226705 (ebook)
Subjects: LCSH: Genetics—Social aspects. | BISAC: SCIENCE / Life Sciences /
    Genetics & Genomics | POLITICAL SCIENCE / Public Policy / Social Policy
Classification: LCC QH438.7 .H39 2021 (print) | LCC QH438.7 (ebook) |
    DDC 304.5—dc23
LC record available at https://lccn.loc.gov/2021012198
LC ebook record available at https://lccn.loc.gov/2021012199

British Library Cataloging-in-Publication Data is available

Editorial: Alison Kalett, Hallie Schaeffer
Production Editorial: Terri O'Prey
Jacket/Cover Design: Karl Spurzem
Production: Danielle Amatucci
Publicity: Sara Henning-Stout, Kate Farquhar-Thomson
Copyeditor: Annie Gottlieb

Jacket art: Shutterstock

This book has been composed in Adobe Text and Gotham

Printed on acid-free paper. ∞

Printed in the United States of America

10 9 8 7 6 5 4 3 2 1

For Jonah and Rowan, my fortune

I used to believe that luck was a thing outside me, a thing that governed only what did and didn't happen to me. . . . Now I think I was wrong. I think my luck was built into me, the keystone that cohered my bones, the golden thread that stitched together the secret tapestries of my DNA.

—TANA FRENCH, *THE WITCH ELM*

CONTENTS

PART I

Taking Genetics Seriously

1

Introduction

In the summer before my son started kindergarten, my mother, suspicious of the Montessori approach I had taken to his preschool education, offered to help him get ready for what she calls "real" school (the kind with desks). I was fairly confident that his transition to kindergarten would go fine, but I nevertheless seized my chance to go on "real" vacation (the kind without small children). Off my children went to spend two weeks with their grandmother, while I spent two weeks on a beach.

My mother used to be a schoolteacher. A speech pathologist by training, she worked in a semi-rural school district in northern Mississippi, where her students often had serious learning disabilities and were always poor. Now that she's retired, the sunroom in her house in Memphis is decorated with posters scavenged from her old classroom: the ABCs, the US presidents, the world's continents, the Pledge of Allegiance. When I returned from vacation, my children could proudly recite: "I pledge allegiance to the Flag of the United States of America, and to the Republic for which it stands, one Nation under God, indivisible, with liberty and justice for all."

On the poster's laminated surface, my mother had used a purple marker to annotate the text of the Pledge of Allegiance with more child-friendly words. Above Republic, she wrote "country." Above liberty, she wrote "freedom." Above justice, she wrote, "being fair."

"Being fair" works admirably well as a kindergarten-friendly definition of justice. As any parent who has seen siblings squabble over a toy can attest, children have a keen sense of fairness and unfairness. If tasked with dividing up some colorful erasers to reward other children for cleaning their rooms, elementary school children will throw away an extra eraser rather than give one child an unequal share.[1]

Even monkeys have a sense of fairness. If two capuchin monkeys are "paid" in cucumber slices for performing a simple task, they will both happily pull levers and munch on their cucumber snacks. Start paying just one monkey in grapes, however, and watch the other monkey throw the cucumber back in the experimenter's face with the indignation of Jesus flipping the tables of the moneychangers.[2]

As human adults, we share with our children and our primate cousins an evolved psychology that is instinctively outraged by unfairness. Right now, such outrage is bubbling all around us, threatening to boil over at any moment. In 2019, the three richest billionaires in the US possessed more wealth than the poorest 50 percent of the country.[3] Like capuchin monkeys being paid in cucumbers when their neighbor is being paid in grapes, many of us look at the inequalities in our society and think: "This is unfair."

To the Educated Go the Spoils

Life, of course, is unfair—including how long one's life is. Across many species, from rodents to rabbits to primates, animals who are higher in the pecking order of social hierarchy live longer and healthier lives.[4] In the United States, the richest men live, on average, 15 years longer than the poorest, who have life expectancies at age 40 similar to men in Sudan and Pakistan.[5] In my lab's research, we found that children growing up in low-income families and neighborhoods show epigenetic signs of faster biological aging when they are as young as 8 years old.[6] It might be easier for a camel to pass through the eye of a needle than a rich man to enter the gates of

Heaven, but the rich man has the consolation of being able to fore-stall judgment day.

These income inequalities are inextricable from inequalities in education. Even before the novel coronavirus pandemic, life spans for White[7] Americans without a college degree were actually get-ting shorter.[8] This historically unusual decline in life span, unique among high-income countries, was driven by an epidemic of "deaths of despair," including overdoses from opioid drugs, complications from alcoholism, and suicides.[9] The coronavirus pandemic made things worse. In the US, people with a college education are more likely to have jobs that can be done remotely from home, where they are more protected from exposure to a virus—and more protected from layoffs.[10]

In addition to living longer and healthier lives, the educated also make more money. In the past forty years, the top 0.1 percent of Americans have seen their incomes increase by more than 400 percent, but men without a college degree haven't seen any increase in real wages since the 1960s.[11] *The 1960s.* Think about how much has changed since then: We have put a man on the moon; we have fought wars in Vietnam and Kuwait and Afghanistan and Iraq and Yemen; we invented the internet and DNA editing; and in all that time, American men who didn't get past high school haven't gotten a raise.

When economists talk about the relationship between income and education, they use the term "skills premium," which is the ratio of wages for "skilled" workers, meaning ones that have a col-lege degree, to "unskilled" workers, meaning ones who don't. This conception of "skill" leaves out tradespersons, like electricians or plumbers, who can have lengthy and specialized training via appren-ticeship rather than college. And anyone who has ever worked an allegedly "unskilled" job like waiting tables will rightly scoff at the idea that such labor doesn't require skill. Working in food service, for instance, involves supplying emotional energy to other people, displaying feelings in the service of how other people feel.[12] The lan-guage of "unskilled" vs. "skilled" workers can reflect what the writer

Freddie deBoer has called "the cult of the smart":[13] the tendency to fetishize the skills that are cultivated and selected for in formal education as inherently more valuable than all other skills (e.g., manual dexterity, physical strength, emotional attunement).

In the United States, the magnitude of the "skills premium" in wages has been increasing since the 1970s, and as of 2018, workers with a bachelor's degree earned, on average, 1.7 times the wage of those who had completed only high school.[14] People who lack an even more basic marker of "skill"—a high school diploma—fare even worse. This is not a trivial number of people: The high school graduation rate has barely budged since the 1980s, and about 1 in 4 high school students will not receive a diploma.[15]

The skills premium is about what an individual worker earns in wages. But many people don't work, and many people don't live alone. Differences in the composition of households further exacerbate inequality. Now more than ever, college-educated people marry and mate with other college-educated people, concentrating high earnings potential within a single household.[16] At the same time, rates of solo parenting and total fertility rates are higher for women with less education.[17] In 2016, 59 percent of births to women with only a high school degree were non-marital, compared to 10 percent of births to women with a bachelor's degree or higher. So, non-college-educated women earn less money, have more mouths to feed, and are less likely to have anyone else in the house to help them pull it off.

These social inequalities leave their mark psychologically. People with lower incomes report feeling more worry, stress, and sadness, and less happiness, than people making more money.[18] They are more immiserated by negative events both large (divorce) and small (headache). They even enjoy their weekends less. On the other hand, global life satisfaction—"my life is the best possible life for me"— goes up with income, even among high earners.

Given the myriad ways that people's lives can end up unequal, philosophers have debated which one is the most important: Some consider equality of monetary resources to be the main thing to worry about. Some consider money simply a means to happiness or well-being. Some refuse to settle on a single currency of justice. Similarly, social scientists tend to study the type of inequality that is the focus of their disciplinary training. For example, economists are particularly likely to study differences in income and wealth, whereas psychologists are more likely to study differences in cognitive abilities and emotions. There is no single best place to start when considering the tangled nest of inequalities between people. But in the US today, whether one is a member of the "haves" or the "have-nots" is increasingly a matter of whether or not one has a college degree. If we can understand why some people go further in school than others do, it will illuminate our understanding of multiple inequalities in people's lives.

Two Lotteries of Birth

People end up with very different levels of education and wealth and health and happiness and life itself. Are these inequalities *fair*? In the pandemic summer of 2020, Jeff Bezos added $13 billion to his fortune in a single day,[19] while 32 percent of US households were unable to make their housing payment.[20] Looking at the juxtaposition, I feel a bubbling disgust; the inequality seems obscene. But opinions differ.

When discussing whether inequalities are fair or unfair, one of the few ideological commitments that Americans broadly claim to share (or at least pay lip service to) is a commitment to the idea of "equality of opportunity." This phrase can have multiple meanings: What, exactly, counts as real "opportunity," and what does it take to make sure it's equalized?[21] But, generally, the idea is that all people, regardless of the circumstances of their birth, should have the same opportunities to lead a long and healthy and satisfying life.

Through the lens of "equality of opportunity," it is not strictly the size or scale of inequalities per se that is evidence that society is unfair. Rather, it is that those inequalities are tied to the social class of a child's parents, or to other circumstances of birth that are beyond the child's control. Whether one is born to rich parents or poor ones, to educated or uneducated ones, to married or unmarried ones, whether you go home from the hospital to a clean and cohesive neighborhood or a dirty and chaotic one—these are accidents of birth. A society characterized by equality of opportunity is one in which these accidents of birth do not determine a person's fate in life.

From the perspective of equality of opportunity, several statistics about American inequality are damning. On the left side of figure 1.1, I've illustrated one such statistic: how rates of college completion differ by family income. It's a familiar story. In 2018, young adults whose families were in the top quarter of the income distribution were nearly four times more likely to have completed college than those whose families were in the bottom quarter of the income distribution: 62 percent of the richest Americans had a bachelor's degree by age 24, compared with 16 percent of the poorest Americans.

It is important to remember that these data are correlational. We don't know, from this data alone, why families with more money have children who are more likely to complete college, or whether simply giving people more money would cause their children to go further in school.[22]

Yet, in public debates and academic papers about inequality, two things are taken for granted about such statistics. First, data on the relationship between the social and environmental conditions of a child's birth and his or her eventual life outcomes are agreed to be *scientifically useful*. Researchers who hoped to understand patterns of social inequality in a country, but who had no information about the social circumstances into which people were born, would be incredibly hampered. Lifelong careers are devoted to trying to understand *why*, exactly, high-income children go further in school, and trying to design policies and interventions to close income gaps in education.[23] Second, such statistics are agreed to be *morally relevant*. For many people, the distinction they make between inequalities that are fair and those that are unfair is that unfair inequalities are those tied to accidents of birth over which a person has no control, like being born into conditions of privilege or penury.

But there is another accident of birth that is also correlated with inequalities in adult outcomes: not the social conditions into which you are born, but the genes with which you are born.

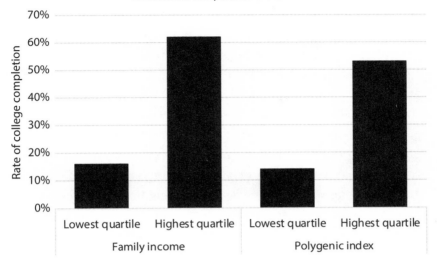

FIGURE 1.1. Inequalities in rates of college completion in the US based on differences in family income versus differences in measured genetics. Data on college completion by income drawn from Margaret W. Cahalan et al., *Indicators of Higher Education Equity in the United States: 2020 Historical Trend Report* (Washington, DC: The Pell Institute for the Study of Opportunity in Higher Education, Council for Opportunity in Education (COE), and Alliance for Higher Education and Democracy of the University of Pennsylvania (PennAHEAD), 2020), https://eric .ed.gov/?id=ED606010. Data on college completion by polygenic index from James J. Lee et al., "Gene Discovery and Polygenic Prediction from a Genome-Wide Association Study of Educational Attainment in 1.1 Million Individuals," *Nature Genetics* 50, no. 8 (August 2018): 1112–21, https://doi.org/10.1038/s41588-018-0147-3; additional analyses courtesy of Robbee Wedow. Polygenic index analyses include only individuals who share genetic ancestry characteristic of people whose recent ancestors all resided in Europe; in the US, these people are very likely to be racially identified as White. The distinction between race and genetic ancestry will be described in more detail in chapter 4.

On the right side of figure 1.1, I have graphed data from a paper in *Nature Genetics*[24], in which researchers created an *education polygenic index* based entirely on which DNA variants people had or didn't have. (I will describe in detail how polygenic indices are calculated in chapter 3.) As we did for family income, we can look at rates of college completion at the lower end versus the upper end of this polygenic index distribution. The story looks much the same: those whose polygenic indices are in the top quarter of the "genetic" distribution were nearly four times more likely to graduate from college than those in the bottom quarter.

The data on family income on the left, despite being correlational, is considered critically important as a starting point for understanding inequality. Social class is recognized as a systemic force that structures who gets more education, and who gets less. The data on family income is also considered by many to be prima facie evidence of unfairness—an inequality that demands to be closed. But what about the data on the right?

In this book, I am going to argue that the data on the right, showing the relationship between measured genes and educational outcomes, is also critically important, both empirically and morally, to understanding social inequality. Like being born to a rich or poor family, being born with a certain set of genetic variants is the outcome of a lottery of birth. You didn't get to pick your parents, and that applies just as much to what they bequeathed you genetically as what they bequeathed you environmentally. And, like social class, the outcome of the genetic lottery is a systemic force that matters for who gets more, and who gets less, of nearly everything we care about in society.

How Genetics Is Perceived

To insist that genetics is, in any way, relevant to understanding education and social inequality is to court disaster. The idea seems dangerous. The idea seems—let's be frank—eugenic. One historian compared scientists who linked genetics with outcomes such as college completion to Germans who were complicit in the Holocaust ("CRISPR's willing executioners").[25] Another colleague once emailed me to say that conducting research on genetics and education made me "no better than being a Holocaust denier." In my experience, many academics hold the conviction that discussing genetic causes of social inequalities is fundamentally a racist, classist, eugenic project.

We also have some insights into how the general public perceives scientists who talk about genetically-caused individual differences—and it's not pretty.

In one social psychology study, participants were asked to read a story about a fictional scientist, Dr. Karlsson.[26] There were two versions of the vignette. In both, the fictional Dr. Karlsson's research program and scientific methods were described in *exactly* the same way. What differed was Dr. Karlsson's results: In one version, participants read that Dr. Karlsson found that genetic causes were weakly associated with performance on a math ability test, accounting for about 4 percent of the variation between people. In the other version, genetic influences were stronger, accounting for 26 percent.

After reading about these research findings, participants were asked how likely it was that Dr. Karlsson would agree with five statements:

1. People's status in society *should* correspond with their natural ability.
2. I believe people and social groups *should* be treated equally, independently of ability.
3. Some people *should* be treated as superior to others, given their hard-wired talent.
4. It's *OK* if society allows some people to have more power and success than others—it's the law of nature.
5. Society *should* strive to level the playing field, to make things just.

These statements were intended to measure "egalitarian" values. The Merriam-Webster definition of *egalitarianism* is "a belief in human equality especially with respect to social, political, and economic affairs; a social philosophy advocating the removal of inequalities among people." When participants read that Dr. Karlsson found evidence for stronger genetic causes of math ability, they perceived him as having less-egalitarian values—as wanting to treat some people as superior to others, as being uninterested in making society more just, as not believing that people should be treated equally.

Furthermore, this study found that a scientist who reported genetic influence on intelligence was also perceived as less objective, more motivated to prove a particular hypothesis, and more likely to hold non-egalitarian beliefs that predated their scientific research career. People who described themselves as politically conservative doubted scientists' objectivity across the board, regardless of the scientists' findings, but people who described themselves as politically liberal were particularly likely to doubt the scientist's objectivity when she reported genetic influences on intelligence.

This study is important because the participants were not scientists or academics with any particular expertise in genetics or mathematics or political philosophy. They were college undergraduates fulfilling a course requirement, or people working from home who wanted to earn some extra money by filling out surveys. The study speaks to how common it is for people, particularly when they have liberal political ideologies, to see empirical statements about how genes *do* influence human behavior as incompatible with moral beliefs about how people *should* be treated equally.

The Enduring Legacy of Eugenics

There are, of course, good reasons why many people perceive genetic findings to be incompatible with social equality. For over 150 years, the science of human heredity has been used to advance racist and classist ideologies, with horrific consequences for people classified as "inferior."

In 1869, Francis Galton—cousin of Charles Darwin and coiner of the term "eugenics"—published his book *Hereditary Genius*.[27] Essentially consisting of hundreds of pages of genealogies, Galton's book aimed to demonstrate that British class structure was generated by the biological inheritance of "eminence." Men with great professional achievements in science, business, and the law descended from other great men. *Hereditary Genius*, along with Galton's subsequent 1889 book *Natural Inheritance*,[28] reframed the study of "heredity" as the study of measurable similarities between relatives[29]—a scientific approach that continues today, including in many of the studies I will describe in this book.

Galton, however, wasn't content merely to document familial resemblance in the form of pedigree tables; he wanted to *quantify*—put a number on—that resemblance. Indeed, quantification was his most enduring enthusiasm; "whenever you can, count" was his slogan.[30] In seeking a mathematical representation of familial resemblance, Galton invented foundational statistical concepts, like the correlation coefficient. But alongside his statistical developments, he also speculated about how heredity could and should be manipulated in humans. In a footnote published in 1883, Galton introduced the new word "eugenics" to "express the science of improving stock," the aim of which was "to give to more suitable races or strains of blood a better chance of prevailing speedily over the less suitable."[31] From the very beginning, then, the nascent science of statistics, and the application of statistics to study patterns of familial resemblance, were entangled with beliefs about racial superiority and with proposals to intervene in human reproduction for the goal of species betterment.

When he died in 1911, Galton bequeathed money to University College London for a Galton Eugenics Professorship, a position that was given to his protégé, Karl Pearson, who was also the head of the newly created Department of Applied Statistics.[32] In his role, Pearson continued to make foundational contributions to statistical methods that are now routinely used in every branch of science and medicine. His research activities were cloaked in a language of neutrality: "We of the Galton laboratory have no axes to grind. We gain nothing, and we lose nothing, by the establishment of the truth." Yet Pearson's political agenda was anything but neutral. Brandishing statistics about familial correlations for "mental characteristics" (such as teacher ratings of academic ability), Pearson argued that progressive-era social reforms, like the expansion of education, were useless. He also opposed labor protections, such as prohibitions on child labor, the minimum wage, and the eight-hour workday, on the grounds that these reforms encouraged reproduction among "incapables."[33]

In the United States, Galton and Pearson's enthusiasm for quantitative studies of family pedigree data was mirrored in the work of Charles B. Davenport, who established a Eugenics Record Office at Cold Spring Harbor on Long Island, New York. In 1910, Davenport appointed Harry H. Laughlin as the Office's superintendent, thus empowering perhaps the most effective proponent of eugenic legislation in American history.

Almost immediately after beginning his post, Laughlin began research for his book, *Eugenical Sterilization in the United States*,[34] which was eventually published in 1922. Citing legal precedents such as compulsory vaccination and quarantine, Laughlin's book argued in support of "the right of the state to limit human reproduction in the interests of race betterment." The book culminated in text for a "Model Eugenical Sterilization Law," to be adapted by state legislatures interested in preventing "the procreation of persons socially inadequate from defective inheritance." "Socially inadequate" persons were defined as anyone who "fails chronically . . . to maintain himself or herself as a useful member of the organized social life of the state," as well as the "feeble-minded," insane, criminally delinquent, epileptic, alcoholic, syphilitic, blind, deaf, crippled, orphaned, homeless, and "tramps and paupers." In 1924, the state of Virginia passed a Sterilization Act that used language directly from Laughlin's model law.[35]

Eugenicists eager to establish the constitutionality of Virginia's Eugenical Sterilization Act quickly found an ideal test case in Carrie Buck, whose own mother, Emma, had syphilis, and who had given birth to a daughter, Vivian, while unmarried, after being raped by her foster parent's nephew.[36] Writing for the majority in *Buck v. Bell*, Supreme Court justice Oliver Wendell Holmes upheld the Virginia statute with an infamous pronouncement on the Buck family: "Three generations of imbeciles is enough." After the *Buck v. Bell* decision, and continuing until 1972, more than 8,000 Virginians were sterilized, and around 60,000 Americans were sterilized as other states followed Virginia's example.[37]

Still, the pace of sterilization was too slow to satisfy the most zealous proponents of eugenics. When Germany passed its own version of Laughlin's model law, soon after Hitler gained power in 1933, American eugenicists urged the expansion of sterilization programs here. "The Germans are beating us at our own game," bemoaned Joseph DeJarnette, a plantation-born son of the Confederacy, who had testified against Carrie Buck in *Buck v. Bell* and who oversaw over 1,000 sterilizations as the director of Western State Hospital in Staunton, Virginia.[38]

In 1935, the Nazi government passed the Nuremberg Laws, prohibiting marriage between Jews and non-Jewish Germans, and stripping Jews, Roma, and other groups of legal rights and citizenship. That year, Laughlin wrote to his Nazi colleague, Eugen Fischer, whose work on the "problem of miscegenation" had provided an ideological foundation for the Nuremberg Laws.[39] The goal of Laughlin's letter to Fischer was to introduce him to Wickliffe Preston Draper, a textile magnate and eugenics enthusiast who would be soon traveling to Berlin to attend a Nazi conference on "race hygiene."[40]

Upon his return to the US, Draper worked with Laughlin to establish the Pioneer Fund, which was incorporated in 1937 and still exists today. Named in honor of the "pioneer" families who originally settled the American colonies, the fund aimed to promote research on human heredity and "the problems of race betterment." One of its first activities was to distribute a Nazi propaganda film on sterilization, *Erbkrank,* which had received special acknowledgment from Hitler himself.[41]

We can draw a direct line, both financially and ideologically, from these eugenicists of the early twentieth century to the white supremacists of today. Consider, for example, Jared Taylor, a self-described "race realist" who thinks that Black Americans are incapable of "any sort of civilization"—and a recent recipient of Pioneer Fund money.[42] Continuing in the ideological tradition of Pearson and Laughlin, Taylor embraces genetics as a rhetorical weapon against the goals of social and political equality. His review of *Blueprint,* a book by the behavioral geneticist Robert Plomin (whose work I will describe in this book), proclaimed that new developments in genetics would sound the death knell for social justice: "if [these] scientific findings were broadly accepted, they would destroy the basis for the entire egalitarian enterprise of the last 60 or so years."[43]

In 2017, white supremacists converged in Charlottesville for the "Unite the Right" rally.[44] Men in khakis waved swastika flags and chanted "Jews will not replace us" as they marched through the town where Carrie Buck is buried—a grim reminder that the demented ideology of "racial purity" connecting Jim Crow Virginia and Nazi Germany, an ideology that also had grisly consequences for poor Whites like Buck, has never fully gone away.

Genetics and Egalitarianism: A Preview

In the century and a half since the publication of *Hereditary Genius*, geneticists have identified the physical substance of heredity, discovered the double-helix structure of DNA, cloned a sheep, sequenced the genomes of anatomically modern humans and of Neanderthals, created three-parent embryos, and pioneered CRISPR-Cas9 technology to edit the DNA code directly. Yet, in all that time, how people make sense of the relationship between genetic differences and social inequalities has barely budged from Galton's original formulation: empirical claims ("people differ genetically, which causes physical, psychological, and behavioral differences") are mixed together with moral oughts ("some people should be treated as superior to others"), with potentially horrible consequences.

What I am aiming to do in this book is re-envision the relationship between genetic science and equality. Can we peel apart human behavioral genetics, beginning with Galton's observations and continuing to modern genetic studies of intelligence and educational attainment, from the racist, classist, and eugenicist ideologies it has been entwined with for decades? Can we imagine a new synthesis? And can this new synthesis broaden our understanding of what equality looks like and how to achieve it?

To begin to convey how we can reimagine the relationship between genetics and egalitarianism, it will help here to describe where I diverge from a book in the Galtonian tradition—*The Bell Curve*, by Richard Herrnstein and Charles Murray.[45] The title of *The Bell Curve* is a nod to Galton's statistical preoccupation, the observation that plotting the population frequency of different values of human traits results in a bell-shaped "normal" distribution with particular mathematical properties. The subtitle (*Intelligence and Class Structure in American Life*) is a nod to Galton's social preoccupation, the question of how class differences reflected genetic inheritance.

Instead of "eminence," Herrnstein and Murray focused on intelligence, as measured by standardized tests of abstract reasoning skills. Like Herrnstein and Murray (and like the vast majority of psychological scientists), I also believe that intelligence tests measure an aspect of a person's psychology that is relevant for their success in contemporary educational systems and labor markets, that twin studies tell us something meaningful about the genetic causes of individual differences between people, and that intelligence is heritable (a terribly misunderstood concept that I will explain in detail in chapter 6). Given these similarities, comparisons between this book and *The Bell Curve*, along with Herrnstein's earlier 1973 book on IQ and meritocracy,[46] are unavoidable. Briefly enumerating the differences between us here, therefore, has the advantage not just of pre-empting misunderstandings but also of foreshadowing the arguments I will advance throughout this book.

Here, I will argue that the science of human individual differences is entirely compatible with a full-throated egalitarianism. The final section of *The Bell Curve* flirts with the idea that genetics could be used to bolster egalitarian arguments for greater economic equality: "Why should [someone] be penalized in his income and social status? . . . We could grant that it is a matter not of just deserts but of economic pragmatism about how to produce compensating benefits for the least advantaged members of society."

There are two big ideas crammed into these few sentences: (1) that people do not *deserve* economic disadvantages simply because they happened to inherit a particular combination of DNA, and (2) that society should be organized so that it benefits the least advantaged members of society. It's disorienting to come across these ideas in *The Bell Curve*, because they sound like they come straight out of a very different book: *A Theory of Justice*, by the egalitarian political philosopher John Rawls.

In *A Theory of Justice*, Rawls used the metaphor of the "natural lottery" to describe how people differ in their initial positions in life. As I'll describe in chapter 2, a lottery is a perfect metaphor for describing genetic inheritance: the genome of every person is the outcome of nature's Powerball.

Rawls then devotes several hundred pages to considering how a just society should be arranged, given that people do differ in the outcome of two lotteries of birth, the natural and the social. Far from seeing differences between people in their "natural abilities" as justifying inequalities, Rawls decried the injustice of societies that were structured according to the "arbitrariness found in nature." His principles of justice led him to argue that inequalities that stemmed from the natural lottery were acceptable only if they worked to the benefit of the least advantaged in society. In Rawls's view, taking biological differences between people seriously did not undermine the case for egalitarianism; it was part of the reasoning that led to him to advocate for a *more* equal society.

The Bell Curve, with its fleeting reference to Rawlsian ideas, pointed faintly at a new way of talking about genetics and social equality. But after their tantalizing half-page dalliance with egalitarianism, Herrnstein and Murray retreat to a profound *in*egalitarianism, complaining that "it has become objectionable to say that some people are *superior* to other people. . . . We are comfortable with the idea that some things are *better* than others—not just according to our subjective point of view but according to enduring standards of merit and *inferiority*" (emphasis added). After 500 pages, it's clear what sort of things—and what type of people—they consider better. According to them, to score higher on IQ tests is to be superior; to be White is to be superior; to be higher class is to be superior. Indeed, they describe economic productivity ("putting more into the world than [one] take[s] out") as "basic to human dignity."

Compare their slick confidence that some people are superior to other people with the definition of inegalitarianism provided by the political philosopher Elizabeth Anderson:[47]

> Inegalitarianism asserted the justice or necessity of basing social order on a hierarchy of human beings, ranked according to intrinsic worth. Inequality referred not so much to distributions of goods as to relations between superior and inferior persons. . . . Such unequal social relations generate, and were thought to justify, inequalities in the distribution of freedoms, resources, and welfare. This is the core of inegalitarian ideologies of racism, sexism, nationalism, caste, class, and eugenics.

In other words, eugenic ideology asserts that there is a hierarchy of superior and inferior human beings, where one's DNA determines one's intrinsic worth and rank in the hierarchy. The social, political, and economic inequalities that proceed from this hierarchy—where the superior get more, and the inferior get less—are, according to eugenic thought, inevitable, natural, just, and necessary.

The standard rejoinder to eugenic ideology has been to emphasize people's genetic sameness. After all, differences between people in their DNA cannot be used to determine their worth and rank if there are no differences. This rhetoric, linking political and economic equality to genetic similarity, is clearly evident in how President Bill Clinton announced that the Human Genome Project had completed its first complete rough draft of the sequence of human DNA.[48] He trumpeted the genetic sameness of humans as an empirical truth that buttressed an egalitarian ideal:

> All of us are created equal, entitled to equal treatment under the law. . . . I believe one of the great truths to emerge from this triumphant expedition inside the human genome is that in genetic terms, all human beings, regardless of race, are more than 99.9 percent the same.

As Clinton said on a different occasion, "mistakes were made," and tying genetic sameness to egalitarian ideals was, I believe, one of Clinton's mistakes. Yes, the genetic differences between any two people are tiny when compared to the long stretches of DNA coiled in every human cell. But these differences loom large when trying to understand why, for example, one child has autism and another doesn't; why one is deaf and another hearing; and—as I will describe in this book—why one child will struggle with school and another will not. Genetic differences between us matter for our lives. They *cause* differences in things we care about. Building a commitment to egalitarianism on our genetic uniformity is building a house on sand.

The biologist J.B.S. Haldane compared Karl Pearson to Christopher Columbus: "His theory of heredity was incorrect in some fundamental respects. So was Columbus's theory of geography. He set out for China, and discovered America."[49] The comparison of Columbus with Pearson and his fellow eugenicists is the right one,

I think. They are similar in the enormity of their theoretical incorrectness, in the enormity of the violence and harm they brought to innocent people—and in the enormousness of what they discovered. Knowing what we know now, we cannot pretend that the continent of America does not exist. Knowing what we know now, we cannot pretend that genetics do not matter. Instead, we must carefully scrape away the eugenicists' scientific and ideological errors, and we must articulate how the science of heredity can be understood in an egalitarian framework.

In this book, I will argue that it is not eugenic to say that people differ genetically. Nor is it eugenic to say that genetic differences between people cause some people to develop certain skills and functionings more easily. Nor is it eugenic for social scientists to document the ways in which educational systems and labor markets and financial markets reward people, financially and otherwise, for a particular, historically and culturally contingent set of genetically influenced talents and abilities. What *is* eugenic is attaching notions of inherent inferiority and superiority, of a hierarchical ranking or natural order of humans, to human individual differences, and to the inheritance of genetic variants that shape these individual differences. What *is* eugenic is developing and implementing policies that create or entrench inequalities between people in their resources, freedoms, and welfare on the basis of a morally arbitrary distribution of genetic variants.

The anti-eugenic project, then, is to (1) understand the role that genetic luck plays in shaping our bodies and brains, (2) document how our current educational systems and labor markets and financial markets reward people with certain types of bodies and brains (but not other types of brains and bodies), and (3) reimagine how those systems could be transformed to the inclusion of everyone, regardless of the outcome of the genetic lottery. As the philosopher Roberto Mangabeira Unger wrote, "Society is made and imagined . . . it is a human artifact rather than the expression of an underlying natural order."[50] This book views the understanding of the natural world, in the form of genetics, as an ally rather than an enemy in the remaking and reimagining of society.

Why We Need a New Synthesis

That genetics would be useful at all for advancing the goals of social equality is a claim that is frequently met with skepticism. The potential dangers of eugenics loom large in the imagination. The potential benefits of connecting genetics to social inequalities, on the other hand, might seem slim. Even if a new synthesis of genetics and egalitarianism is possible, why take the risk? Given the dark legacy of eugenics in America, it might feel overly optimistic, even naïve, to imagine that genetic research could ever be understood and used in a new way.

What is missing from this consideration of risks and benefits, however, are the risks of continuing the status quo, where understanding how genetic differences between individuals shape social inequalities is widely considered, by both academics and the lay public, to be taboo. This status quo is no longer tenable.

As I will explain in chapter 9, the widespread tendency to ignore the existence of genetic differences between people has hobbled scientific progress in psychology, education, and other branches of the social sciences.[51] As a result, we have been much less successful at understanding human development and at intervening to improve human lives than we could be. There is not an infinite supply of political will and resources to spend on improving people's lives; there is no time and money to waste on solutions that won't work. As the sociologist Susan Mayer said, "if you want to help [people], you have to *really* know what help they need. You can't just think you have the solution"[52] (emphasis added). If social scientists are collectively going to rise to the challenge of improving people's lives, we cannot afford to ignore a fundamental fact about human nature: that people are not born the same.

Ignoring genetic differences between people also leaves an interpretive vacuum that political extremists are all too happy to fill. Jared Taylor is not the only extremist to retain an interest in genetics. As the geneticists Jedidiah Carlson and Kelley Harris summarized, "members and affiliates of white nationalist movements are voracious consumers of scientific research."[53] Both journalists and scientists have sounded the alarm about how genetics research was dissected on white supremacist websites like Stormfront (motto: "White Pride Worldwide"),[54] but Carlson and Harris were able to put hard numbers on the phenomenon by analyzing data on how social media users shared working papers that scientists had posted to *bioRxiv*. Their analysis showed that papers on genetics are particularly popular among white nationalists.

I've seen this phenomenon play out with my own work. Take, for example, a paper I co-authored on how genetic differences are related to what economists have called "non-cognitive skills" related to success in formal education. (I'll explain this paper in more detail in chapter 7).[55] Carlson and Harris's analysis found that five out of six of the biggest Twitter audiences for our paper were people who appeared, from the terms used in their bios and usernames, to be academics in psychology, economics, sociology, genomics, and medicine (figure 1.2). The sixth audience, though, comprised Twitter users whose bios included terms like "white," "nationalist," and the green frog emoji, an image that can be used as a hate symbol in anti-Semitic and white supremacist communities.[56]

This is a dangerous phenomenon. We are living in a golden age of genetic research, with new technologies permitting the easy collection of genetic data from millions upon millions of people and the rapid development of new statistical methodologies for analyzing it. But it is not enough to just produce new genetic knowledge. As this research leaves the ivory tower and disseminates through the public, it is essential for scientists and the public to grapple with what this research *means* about human identity and equality. Far too often, however, this essential task of meaning-making is being abdicated to the most extreme and hate-filled voices. As Eric Turkheimer, Dick Nisbett, and I warned:[57]

> If people with progressive political values, who reject claims of genetic determinism and pseudoscientific racialist speculation, abdicate their responsibility to engage with the science of human abilities and the genetics of human behavior, the field will come to be dominated by those who do not share those values.

Twitter keywords for top 6 audience segments

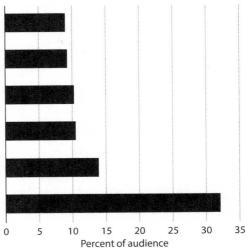

phd, student, genetics, genomics, research, biology, university, science, lab, scientist, postdoc, bioinformatics, biologist, data, molecular, researcher, cancer, fellow, candidate, professor, computational, studying, human, 🧬,...

health, md, medical, healthcare, medicine, care, research, dr, phd, public, physician, director, professor, science, clinical, author, education, family, nutrition, patient, news, researcher, passionate, advocate, services,...

📢, 💟, #maga, ♥, white, nationalist, american, trump, conservative, vida, 🙏, world, god, christian, people, america, ✂, free, truth, 📢, amo, media, news, proud, 🇺🇸, time, country, ✖, music, catholic

research, professor, health, phd, sociology, university, policy, sociologist, science, researcher, prof, assistant, student, data, public, inequality, fellow, population, family, demography, education, associate, political,...

economics, phd, economist, professor, research, development, student, university, policy, assistant, econ, health, candidate, public, education, data, fellow, political, economic, prof, associate, labor, science, researcher,...

phd, research, psychology, genetics, science, university, health, student, professor, psychologist, researcher, neuroscience, cognitive, mental, clinical, dr, brain, scientist, fellow, human, postdoc, assistant, data, studying,...

Percent of audience

FIGURE 1.2. Top 6 largest social media audiences for scientific paper on genetics and non-cognitive skills. Audience analysis methods reported in Jedidiah Carlson and Kelley Harris, "Quantifying and Contextualizing the Impact of bioRxiv Preprints through Automated Social Media Audience Segmentation," *PLOS Biology* 18, no. 9 (September 22, 2020): e3000860, https://doi.org/10.1371/journal.pbio.3000860. Audiences are presented for preprint of Perline Demange et al., "Investigating the Genetic Architecture of Noncognitive Skills Using GWAS-by-Subtraction," *Nature Genetics* 53, no. 1 (January 2021): 35–44, https://doi.org/10.1038/s41588-020-00754-2.

The Goals of This Book

What, then, *does* the science of human abilities and the genetics of human behavior mean for social equality? To address the question, this book proceeds in two general parts. In the first part, I hope to convince you that genetics do, in fact, matter for understanding social inequality. Common counter-arguments to the idea that genetics matter include the ideas that twin studies are hopelessly flawed, that heritability estimates are useless, that associations with measured DNA are just correlations but don't provide any evidence that genes are causal, or that genes might be causal but it doesn't matter if they are if we don't know the mechanism. All of these ideas falter under closer examination, but in order to explain why, it will be necessary to dive into some methodological details of how behavioral genetics research is done, and into some philosophy of science about what those methods are accomplishing.

In chapter 2, I begin by explaining my metaphor of the genetic lottery in more detail, bringing in some biological and statistical concepts, such as genetic recombination, polygenic inheritance, and the normal distribution. Here, and throughout the book, I focus on genetic differences between people that occur because of *chance*, i.e., through the natural lottery of genetic inheritance, rather than because of *choice*, such as through pre-implantation genetic diagnosis or other reproductive technologies.[58]

Next, in chapter 3, I explain common methods for testing how genetic differences between people are associated with differences in their life outcomes, in particular genome-wide association studies and polygenic index studies. Chapter 4 then explains why the results of genome-wide association studies cannot tell us about the causes of *group* differences, particularly differences between racial groups. The unceasing parade of books and articles about "innate" racial differences have been sound and fury signifying nothing. Rather, genetic research on social inequalities, both twin research and research with measured DNA, has focused almost entirely on understanding *individual* differences among people whose recent genetic ancestry is exclusively European[59] and who are overwhelmingly likely to identify as White.

This narrowing of scope provides an essential qualification for all of the empirical results that I describe in the book. Genetic research on social and behavioral phenotypes, with its current focus on people of European genetic ancestry, cannot meaningfully inform our *scientific* understanding of social inequalities between racial and ethnic groups. However, as I describe in chapter 4, our consideration of why people return, time and time again, to the scientifically empty question of genetic racial differences reveals how genetic explanations are used to waive people's social responsibility for enacting change. Considering genetics as an absolution for social responsibility is a false pretext that must be dismantled, regardless of how genes are distributed within or between socially constructed racial groups.

With the distinction between group differences and individual differences in mind, chapter 5 begins to address an essential question about the results of genome-wide association studies and polygenic index studies: Are these studies telling us about genetic *causes*? In order to address this question, I step back and address a more general question first, which is, "What makes something a cause?" Chapter 6 applies this clarity regarding what a cause is (and what it isn't) to understanding the results of genome-wide association studies and heritability studies. Here, too, I review the wealth of evidence showing that genes cause important life outcomes, including educational attainment. Chapter 7 concludes the first half of the book by describing what we know about the mechanisms linking genes and education.

In the second half of the book, I consider what we should do with the knowledge that genetics matter for understanding social inequality. Once we throw away the eugenic formulation that genetic differences form the basis of a hierarchy of innately superior and inferior humans, what is left? In chapters 8 and 9, I consider how understanding genetic differences between people can improve our efforts to change the world through social policy and intervention. In chapter 10, I consider why people are motivated to reject information about genetic causes of human behavior, and how considering genes as a source of luck in people's lives might actually reduce the blame that is heaped on the heads of people who have been "unsuccessful" educationally and economically. In chapter 11, I consider why genetic influences on intelligence test scores and educational outcomes, in particular, are difficult to peel apart from notions of human inferiority and superiority, and compare how we view genetic research on these aspects of human psychology with how we view genetic research on other traits, such as deafness or autism. Finally, in chapter 12, I describe five principles for anti-eugenic science and policy.

Throughout the book, I will not attempt to hide my own left-leaning political sympathies. But my earnest hope is that even readers with politics very different from mine will be convinced that the questions I ponder here are important, even if you vehemently disagree with the answers I suggest. I invite my conservative readers to remember that justice was an idea that also preoccupied the ancient Greeks, the authors of the Bible, and the Founding Fathers. How are we to "do justly," as the prophet Micah exhorted, in a time of accelerating technological change and burgeoning genetic knowledge? I believe this is a question of consequence for us all, regardless of partisanship.

It is audacious to write a book about equality. My own expertise and scholarship, as a psychologist and behavioral geneticist, is in the genetics of human behavior in childhood and adolescence. Theories of equality rarely talk about genes. Theories of equality do, however, talk about skill, talent, ability, endowments, capabilities, ambition, competition, merit, luck, innateness, chance, and opportunity. And, as I hope to show in this book, the field of behavioral genetics has quite a lot to say about all of these things, although what, precisely, genetics can (and cannot) tell us is a good deal more complicated than it might first appear.

2

The Genetic Lottery

The most glamorous person in my daughter's life is an eight-year-old girl named Kyle. She has swishy, waist-length hair pulled back with a sparkly bow headband. She has an extensive collection of *Frozen* dolls. And, most alluringly, Kyle has a trampoline in her front yard.

The trampoline is part of the playscape that Kyle's mom installed the year that her twin brother, Ezra, had brain surgery. Ezra has autism and epilepsy. Most people don't know that children with autism are more likely to have a seizure disorder.[1] I didn't, until I lived next door to Kyle and Ezra, even though I was trained as a clinical psychologist and run a child development research lab. Children with autism who also have intellectual disability, which is clinically defined as having an IQ score less than 70, are particularly vulnerable: over 20 percent also have epilepsy.

When he was four, Ezra's seizures rapidly became so frequent and debilitating that he had a vagus nerve stimulator implanted, a type of pacemaker for his brain. He is still on a strict high-fat, low-carbohydrate ketogenic diet to control his seizure activity.[2] Ezra's mom is an accomplished academic who also makes an excellent ketogenic-compliant chocolate birthday cake.

American parents of children with a diagnosis like autism or intellectual disability almost inevitably come across a 1980s essay called "Welcome to Holland."[3] The premise is that parents of children with special needs are like travelers who planned a one-way trip to Italy. They learned to say *ciao* and looked forward to seeing Michelangelo's *David,* but when their plane lands, the stewardess announces they have landed in Holland instead. There's no leaving Holland. Some parents find this metaphor comforting: "Holland has tulips. Holland even has Rembrandts." Other parents find it infuriating: "I'm tired of Holland and want to go home" is the title of a post on one mother's blog.[4] I've yet to ask a Dutch person what they think of the fact that Americans use their country as a metaphor for parenting a child with a serious disability.

Because Ezra has a twin sister, it's poignantly easy to envision what their family would look like if they had landed in Italy, as planned, instead of in Holland. Kyle bounces on her trampoline with lithe grace and makes easy conversation with adults. Ezra has regressed in the years since his family first moved next door. His speech and social interest have withered; his gait has stiffened. Twins fascinate us because of their sameness, but also because of their differences. Kyle is not Ezra's mirror; she is his counterfactual, his *what if.* And the *what if*s don't stop there. Kyle and Ezra can be compared not only to each other, but also to their unborn triplet, who died in utero.

Although the precise causes of an individual case of pregnancy loss or autism are usually unknown, we can speculate about how genetic differences between Kyle, Ezra, and their unnamed triplet might have shaped their diverging lives. Approximately half of first-trimester miscarriages are due to genetic abnormalities.[5] As much as 90 percent of the variation in vulnerability to autism is due to genetic differences between people. Death before birth; an ostensibly normal infancy followed by a regression into silence; a thriving, bouncy, chatty child—siblings can have dramatically different fates, despite their shared parentage.

A study by psychologists at the University of Minnesota asked people to estimate how much they thought genetic factors "contribute

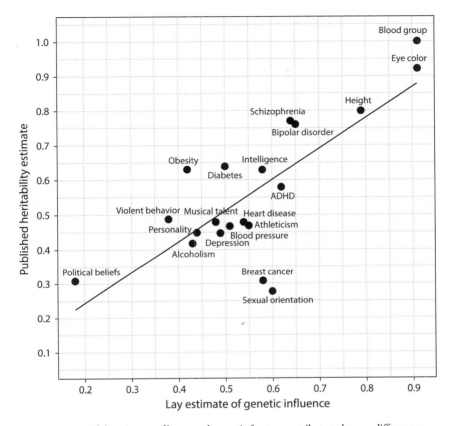

FIGURE 2.1. People's estimates of how much genetic factors contribute to human differences (horizontal axis) versus scientific estimates of heritability from twin studies (vertical axis). The correspondence between lay estimates and scientific estimates is r = .77. Figure reprinted by permission of Springer Nature from Emily A. Willoughby et al., "Free Will, Determinism, and Intuitive Judgments about the Heritability of Behavior," *Behavior Genetics* 49, no. 2 (March 2019): 136–53, https://doi.org/10.1007/s10519-018-9931-1.

to differences among people" in things like their eye color and depression and personality (figure 2.1). Those estimates were then compared to the scientific consensus about the *heritability* of an outcome, as estimated in twin studies—that is, studies that compare how similar identical twins are to how similar fraternal twins are for some characteristic.[6] I will return to the definition of heritability and the details of twin studies in chapter 6, but for now, I just want to point out that lay people's estimates of how much genetics contribute to differences between people match heritability estimates from twin studies fairly closely. And one group of people had intuitions that tracked twin heritabilities especially closely: mothers of multiple children.

This result about moms' accuracy makes sense. Mothers of multiple children have a front-row seat to watch human differences unfold. My own children, although the contrasts between them are not as stark as between Kyle and Ezra, seem to me as different as chalk and cheese. From the moment their second babies come into this world, parents of multiple children experience how every developmental milestone feels new the second time around. The idiosyncrasies of each child can be startling.

In the differences between our children, we see hints of the genetic variation hidden in our cells, and in the cells of our partners. (By genetic variation, I mean differences between people in their DNA sequence.) We readily accept that this genetic variation matters for understanding whether our children are tall or short, whether they have blue eyes or brown, even whether or not they will develop autism. It is more complicated to claim that this genetic variation matters for understanding whether our children will succeed in school, will be financially secure, will commit a crime, will feel satisfied with how their lives are going. It is even more complicated to think about how society should grapple with such genetically associated inequalities. But before we can begin to address these complexities, we first need to establish some basics.

To get us started in this chapter, I am going to describe some biological and statistical concepts, such as recombination, polygenic inheritance, and the normal distribution. A good grasp on these ideas is necessary to make sense of the metaphor of a genetic lottery. With these concepts in mind, I will then give a preview of research studies that show us the power of the genetic lottery for shaping life outcomes. These studies provoke scientific questions about their methods and moral questions about how their results should be interpreted, and these questions, in turn, will occupy us for the rest of the book.

We Contain Multitudes

Bacteria don't bother with sex. Rather, they duplicate themselves to form daughter cells that are identical to one another and their parent. We, on the other hand, need to mix our DNA with someone else's in order to have daughters (or sons), and for that we need gametes: sperm and egg cells. *Meiosis* is the process of making sperm or eggs. During meiosis, we remix the DNA that we inherited from our mother and the DNA that we inherited from our father, creating new arrangements of DNA that have never existed before and will never exist again.

A baby girl is born with about 2 million immature eggs nestled in her tiny ovaries, and over her life about 400 of those will mature and be released during ovulation. Boys don't begin to produce sperm until puberty, and then they churn out, on average, 525 billion over their lifetime.[7] For each sperm or egg cell, the meiotic remixing of DNA begins anew. The resulting combinatorial explosion of potential child genotypes from any two parents is mind-boggling: each pair of parents could produce over 70 *trillion* genetically unique offspring.[8] And that's before you take into account the possibility of *de novo* genetic mutations: brand-new genetic changes that arise in the production of gametes. Like a specific 6-ball combination in Powerball, the fact that you have your specific DNA sequence, out of all the possible DNA sequences that could have resulted from the union of your father and your mother, is pure luck. This is what I mean when I say that your *genotype*—your unique sequence of DNA—is the outcome of a genetic lottery.

For example, there is a genetic *variant* in the *CFH* gene, which codes for something called "complement factor H protein." (By variant, I mean there is more than one version of the gene.) I inherited different versions of this *CFH* variant from each of my parents. In one version of my *CFH* gene, my sequence of DNA letters—called nucleotides—contains a cytosine (abbreviated C). In the other version, it contains a thymine (T). When my tiny fetal body was making even tinier eggs, my T version and my C version were separated and packaged off into different eggs—half my eggs have the T version; half have the C version. In my ovaries, I contain multitudes. As a result, my offspring can be different from one another: my son inherited my T version; my daughter, my C version. I reproduced, and genetic differences that were lurking inside my body were made manifest as genetic differences between my offspring.

It's easy for modern people living in high-income, low-fertility countries to underestimate how different children born into the same family can be. We tend to have small family sizes, if we have children at all. About a quarter of American families have just one child.[9] San Francisco has many dogs as it does children.[10] Shrinking families have shrunk our imaginations about the manifold possibilities of reproduction. (And have made family histories of disease a more impoverished marker of what dangers lurk within our genomes.)

But to glimpse the power of genetic differences between siblings, we can look beyond human families to other species that *do* have large family sizes.[11] Take, for instance, cows: just one black-and-white Holstein bull (named Toystory) had over half a million offspring via artificial insemination. Dairy cows have been the target of intensive artificial selection breeding programs for decades now, which have resulted in dramatic changes in the amount of milk yielded from a single animal. A cow who would have been a top 1-in-1,000 milk-producing cow in 1957 would be a totally average cow now. Importantly, the selective breeding of dairy cows works by exploiting the enormous genetic variety that exists *within* each family. The half million offspring of Toystory represent half a million random samplings from his genome, and the ability to select among these offspring to be the parents of the next generation is what has driven up milk yield so spectacularly in such a short period of time.

In addition to showing the extent and power of within-family genetic variation, selective breeding programs also emphasize the importance of thinking about *combinations* of genetic variants, rather than single genes. The rapid increase in dairy cows' milk yield since the 1950s is not primarily due to the introduction of new genetic mutations. Practically all of the genetic power that resulted in *way* more milk in 2019 was already floating around the gene pool in 1957—one variant here, one variant there. What selective breeding allowed the agricultural industry to do was make the genetic variants that increase milk production more common in the cow population, which increases the number of milk-increasing genetic variants that are concentrated in combination in any one single animal.[12]

Thinking about combinations of many genetic variants, which can be concentrated to varying degrees in a single animal, might be unintuitive. If, like me, you first encountered genetics in high school biology, your introduction to genetics was Gregor Mendel and his pea plants. The pea plant characteristics that Mendel worked with (tall versus short, wrinkly versus smooth, green versus yellow) were determined by a single genetic variant. In contrast, the human characteristics we care most about—things like personality and mental disease, sexual behavior and longevity, intelligence test scores and educational attainment—are influenced by many (very, very, very many) genetic variants, each of which contributes only a tiny drop of water to the swimming pool of genes that make a difference. There is no single gene "for" being smart or outgoing or depressed. These outcomes are *polygenic*.

Moreover, Mendel was working with plants that ordinarily "breed true"—plants with green peas yield more plants with green peas. The offspring of true-breeding plants don't have much diversity. It's easy to graft our conceptions of "inheritance" and "heredity" onto our hazy, high school knowledge of genetics to yield the idea that humans, too, breed true, that children are always like their parents. Mendel's tale of pea plants, like the tales we tell ourselves about how we resemble our parents, about how like descends from like, is a tale of continuity, of similarity, of predictability.

But Mendel's tale of pea plants, the tale of true-breeding plants, the tale of continuity and similarity, is not the right tale to tell about free-range humans. The things we value about ourselves, the things we worry about and *kvell* over in our children, are not like being a smooth or wrinkly pea plant. They are not influenced by one genetic variant, and humans do not breed true.

The Normal Distribution

Dairy cows aren't the only species whose reproduction has been revolutionized by technology. Take Sean and his husband, Daniel, who have been steadily saving money to pay for an egg donor, IVF, and a gestational surrogate, so that they can have biological children.[13] In the summer of 2019, they selected an egg donor—a woman they had talked to over Zoom but had never met in person.

Selecting a partner for reproduction is never random, but the process of choosing an egg donor is liberated from the unconscious and unbiddable forces of romantic and sexual attraction that govern modern mating and marriage. In some ways, that's harder. How do you choose? Their egg donor rides motorcycles. Sean lights up when he talks about science, about singing in a choir, about doing logic puzzles as a child, and about the idea of an egg donor who rides motorcycles.

Half of the eggs harvested from the donor will be inseminated by Daniel; half by Sean. They are hoping for 20 fertilized embryos in all—20 potential full- and half-siblings, created in a way that would have been unimaginable throughout most of human history. With six siblings, nearly twenty nieces and nephews, and so many first cousins that he is unable to name them all, Sean might have been content, maybe, not to have biological children. But Daniel is an only child, and the desire to have a child who is flesh of your flesh, blood of your blood, is a desire that can't be easily vanquished. So, they are using assisted reproductive technology to begin a family.

Just as in dairy cows, assisted reproductive technology in humans makes the workings of the genetic lottery, and the scope of within-family genetic differences, plainer. The 10 sperm that will inseminate each of 10 eggs for each man are a tiny sample of the billions of sperm he will produce in his lifetime. The 20 eggs that the egg donor will produce are a slightly larger sample of her pool of mature eggs. The resulting 20 embryos will be different from each other, genetically. But how different?

I caught up with Sean at a workshop on statistical methods in genetics. A few dozen whip-smart PhD students from economics, sociology, and psychology had gathered to hear about new analyses they could do with large genetic data sets. One of Sean's lectures was on creating polygenic indices. Polygenic indices are the human version of "estimated breeding values" (EBV) in agriculture; Toystory was selected to have half a million offspring because of his EBV. A high EBV for milk production indicates that a bull or cow will have offspring that produce, on average, more milk, and a high polygenic index for height indicates that, all other things about the environment being equal, your offspring will be taller.

When people outside of genetics first hear about polygenic indices, their thoughts immediately go to whether they will be useful for making reproductive decisions like the ones that Sean and Daniel face. Which egg donor to use? Which egg to fertilize? Which embryo to implant? But despite being a world leader in the method of polygenic indices, Sean has no plans to use them to select their egg donor or their embryos. Instead, we talk about how far out "into the tails" twenty embryos could get.

By "the tails," I mean the tails of the genetic distribution. In the late 1800s, Francis Galton, who insisted that the insights of his distant cousin, Charles Darwin, were also relevant for understanding the evolution of human behavior, made perhaps his most unambiguously positive intellectual contribution. He invented a device that illustrated how the normal distribution—the familiar bell-curve shape—could be produced by the accumulation of random events.[14]

A Galton board, or quincunx, is a vertical board with interleaved rows of pegs (left side of figure 2.2). Small beads are dropped from the top of the board, and they jostle through the rows of pegs, randomly bouncing to the left or right at each row, finally making their way to one of the slots at the bottom.

Most of the beads end up in the middle slots, because that's where a bead ends up if it bounces to the right about as many times as it bounces to the left. In order to end up on the far left or the far right—to end up in the *tails*—a bead had to bounce right or left every time. Bouncing right instead of left at every row is like flipping a coin a dozen times and having it comes up heads every time. It happens only rarely, but it *can* happen.

The shape that the beads take at the bottom of a quincunx, with most of them stacked up around the center, and progressively fewer beads as you move from the center to the left or right tail, is the bell curve. The bell curve describes the shape of how many different human characteristics are distributed. If I, for instance, measured the heights of 1,000 people and plotted how many people were 5′0″, 5′1″, 5′2″, etc., all the way up to 6′5″, that graph would look bell-curve-shaped. It would look, in statistical parlance, *normal*.

Galton didn't know what DNA was, because it hadn't been discovered yet. But his observations about the statistical distribution of human characteristics didn't, at first glance, seem to fit what Mendel had discovered about the laws of heredity. Mendel's pea plants came in two heights—short and tall. Crossing short and tall plants didn't yield a group of plants where most plants were medium height. Instead, crossing tall plants with short plants yielded offspring that were all tall; crossing them for a second generation yielded tall and short offspring in a 3:1 ratio. It wasn't at all clear how the emerging science of heredity could account for the patterns observed in the emerging science of statistics.

The apparent paradox was resolved by Ronald Fisher, a seminal figure in modern statistics, population genetics, and experimental design, and a eugenicist who advocated for the sterilization of "mental defectives."[15] (Like Kyle and Ezra, whom I introduced at the beginning of this chapter, Fisher also had his what-might-have-been: his twin, born first, was stillborn.[16]) In a famous 1918 paper, "The Correlation between Relatives on the Supposition of Mendelian Inheritance," Fisher showed that Mendelian inheritance will result in a bell curve distribution of outcomes, so long as the outcome in question is influenced by *many* different "Mendelian factors," which we would now call genetic variants.[17]

Let's revisit the question that I asked Sean about his upcoming round of IVF: How far out "into the tails" could twenty embryos get? Each one of the potential embryos is a bead poised at the top of the board. And each row of pegs represents a genetic variant for which either Daniel or Sean is heterozygous, meaning that he has two different versions of a gene. That is, where there is a possibility

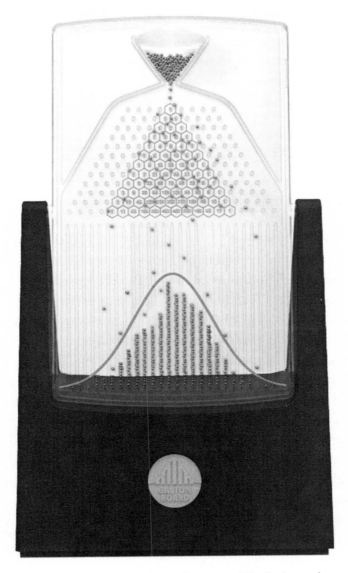

FIGURE 2.2. A Galton board, showing how a normal distribution results from the accumulation of many random events. Photo by Mark Hebner.

that the embryo could inherit either *A* or *a,* the bead could bounce left or right. Bounce left and you get the version that makes you shorter; bounce right and you get the version that makes you taller. Most potential offspring will end up at the bottom of the board in one of the middle slots—they bounced right about as much as they bounced left. They will end up with an about-average number of height-increasing genetic variants. But, still, there's variation; there are differences. Brothers are not the same height. And, once in a blue moon, someone ends up much shorter or much taller than their parents. They end up in the tails of the distribution.

It's Better to Be Lucky than Good

Given that Sean and Daniel and their egg donor are all of relatively average height, it's unlikely that their child will stand as tall as Shawn Bradley, who at 7′6″ tall (~2.3 meters) is one of the tallest basketball players to ever play in the NBA. After sitting next to a geneticist on a plane,[18] Bradley discovered that he was very, very, very far out into the tails of the genetic distribution of height-increasing genetic variants (figure 2.3). Of all the height-increasing variants that he could have inherited, he just happened to get way more than average.[19] Rattling down the quincunx board, Bradley's genome kept bouncing right instead of bouncing left.

How *much* higher than average a number is can be expressed using something called *standard deviation* units. A person who is +1 standard deviation above average in height-increasing genetics has more height-increasing genetic variants than 84 percent of people. A person who is +2 standardized deviations above average has more height-increasing genetic variants than 98 percent of people. Shawn Bradley's was 4.2 standard deviations above average in his number of height-increasing genetic variants, which is higher than 99.999 percent of people. Not the top 1 percent. Not the top 0.1 percent. Not the top 0.01 percent. The top 0.001 percent.

Antonio Regalado, writing for *MIT Tech Review*, quipped that Bradley had "won the jump ball of genetic luck . . . [making] him taller than 99.99999 percent of people."[20] Bradley himself, when reflecting on the genetic inheritance that was critical for a basketball career that resulted in an estimated net worth of $27 million, told the *Wall Street Journal*: "I actually felt pretty lucky and blessed things turned out how they did."[21]

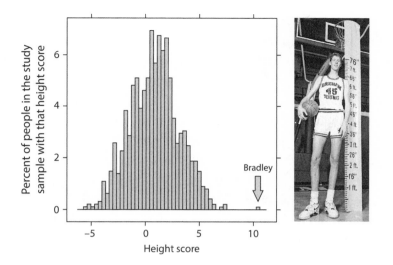

FIGURE 2.3. Height-increasing genetic variants in an individual of extreme height. On the right is a photo of Shawn Bradley next to a ruler showing that he is 7′6″ tall. On the left is the distribution of "genetic scores" (i.e., polygenic indices) constructed from 2,910 genetic variants associated with human height. Mr. Bradley's score was 10.32, whereas the average score in the sample of people being studied was 0.98, with a standard deviation of 2.22. Mr. Bradley's score was 4.2 standard deviations above the mean. Figure adapted from Corinne E. Sexton et al., "Common DNA Variants Accurately Rank an Individual of Extreme Height," *International Journal of Genomics* 2018 (September 4, 2018): 5121540, https://doi.org/10.1155/2018/5121540.

The luck metaphor is apt, but also a new way of thinking about the types of luck that matter in our lives. We ordinarily think of luck as a thing outside ourselves. That time I ended up in the same city with a handsome acquaintance. That time I was driving out of town, feeling dejected at my failed search for an apartment, when I saw someone hammering in a "For Lease" sign in the front yard. That time a cab stopped just inches short of hitting me in a Manhattan crosswalk. We feel luckiest (or unluckiest) when we can clearly imagine what could have come to pass—except it didn't.

But luck is not just something outside of ourselves that happens to us. It is also stitched into us. We are each one in a million—or more literally, one in 70 trillion, which is the number of unique genetic combinations that could have resulted from any two parents. And each of our parents' genomes was a one in 70 trillion event, out of all the possible combinations of *their* parents' DNA, and so on and so forth, back into human history. Each one of our genomes is the ultimate result of generations upon generations of random events that could have gone another way. There is no part of our genome for which we can take credit; no bit of our DNA over which we exerted control.

All of your genome, then, can be considered a form of luck in your life. However, in the effort to understand the *effects* of that genetic luck for behavioral and social outcomes like education or income, scientists often focus on specific parts of the genome—the parts that vary within biological families, where "family" is defined at a distance of just one generation. That is, as I will describe in detail in chapter 6, scientists are often particularly interested in differences between siblings (which DNA did one sibling inherit that their sibling did not?) or the differences between parents and their biological children (which parental DNA did a child inherit versus not inherit?).[22]

This emphasis on studying the genetic lottery as it played out in just one generation—how people differ from their parents and siblings—is necessary because, once you consider multiple generations of genetic differences, those genetic differences are braided together with differences in geography and culture and all the other threads of human history. As a result, understanding what differences between people are caused by genes versus caused by the environments that co-occur with those genes becomes difficult, sometimes to the point of near impossibility.

This entanglement of genetic differences between populations with the environmental and cultural differences between them is called "population stratification." Populations of people differ genetically: for instance, people with East Asian genetic ancestry are more likely to have a certain form of the *ALDH2* gene than people with European genetic ancestry.[23] Populations of people differ culturally: for instance, people raised in East Asian cultures are more likely to use chopsticks than people raised in European cultures. But the co-occurrence between genotype and eating customs is not due to a causal effect of *ALDH2* genetics on chopsticks usage.[24] Subtle forms of population stratification can exist even within groups of people that might seem, at first glance, to be fairly homogenous (e.g., "White British" people in the UK)[25]

In contrast, when we focus on a single generation of the genetic lottery, the science becomes more tractable. The genetic differences between my brother and me, for instance, exist independently of our geography or class or culture. All of my DNA is luck. But studying the part of my DNA that differs from my immediate family is what allows scientists to see the effects of genetic luck more clearly.

Playing for Keeps

The metaphor of genetic lottery captures the randomness inherent in sexual reproduction, but people don't play lotteries just to appreciate the workings of random chance up close. They play for money.

In 2020, three economists—Daniel Barth, Nicholas Papageorge, and Kevin Thom—published a paper in the *Journal of Political Economy* with its thesis in the title: "Genetic Endowments and Wealth Inequality."[26] They argued that genetic differences were relevant not just for individual differences in physical characteristics like height, but also for individual differences in wealth.

Wealth is defined as the total value of your assets (your home, your car, your cash, your retirement savings, your investments and stocks), minus your debts. It's particularly interesting to measure wealth at the time of retirement. By then, a person's wealth reflects a decades-long history of promotions and raises and layoffs and stock market booms and real estate bubbles and inheritances and divorce settlements and student debt payments and alimony payments and putting a kid through college and credit card spending sprees and medical bills. Wealth reflects all the "slings and arrows of outrageous fortune." Including, it turns out, one's genetic fortune.

For their paper, Barth, Papageorge, and Thom focused on a very particular group of Americans: households with one or two adults, where everyone in the house was White, between the ages of 65 and 75, not the same gender, and retired or otherwise not working for pay. That's a pretty narrow slice of America, one that doesn't look like a lot of the country. But even within that relatively homogeneous group, some Americans are much wealthier than others. The bottom 10 percent of people had an average of about $51,000; the top 10 percent of people had over $1.3 million.

In order to measure a person's "genetic endowments," Barth, Papageorge, and Thom used a polygenic index.[27] In the next chapter, I will discuss in much more detail how a polygenic index is constructed. For now, a polygenic index is a single number that adds up how many genetic variants a person has, based on previous research estimating how strongly those genetic variants are related to a measured outcome. So, a polygenic index for height, such as the one used to study the NBA player Shawn Bradley, takes information from previous studies about which DNA variants are correlated with being taller, and uses that information to add up how many "being-taller" variants a person has. In this study on wealth, the investigators were focusing on a polygenic index that summarized information about DNA variants known to be associated with educational attainment (i.e., staying in school for more years), and compared how much wealth people had across different levels of the polygenic index.

Among the White, retired septuagenarians in this study, people who were low on the polygenic index (the first quartile) had, on average, *$475,000* less wealth than people who were high on the polygenic index (the fourth quartile). Another way of expressing that same result is that people who were +1 standard deviation higher in the polygenic index had nearly 25 percent more wealth. Although the polygenic index was constructed based on DNA variants associated with staying in school for longer, people who are high on the polygenic index don't necessarily have more schooling, and their result can't be explained just by the fact that people who have more schooling make more money. Even when comparing people who have the *same* amount of education, a +1 standard deviation increase in their polygenic index was associated with an 8 percent increase in wealth.

But, in their analysis, Barth, Papageorge, and Thom were not comparing siblings; they were comparing different households. This is important, because, as I mentioned previously, some genetic luck is braided together with other differences between families. Could the relationship between genes and wealth be due to this problem of population stratification? For instance, people with higher polygenic indices also are more likely to have parents with higher education. Consequently, people with "luckier" genes also won the social lottery, as they were the beneficiaries of more-advantaged childhood environments. And they were potentially the beneficiaries of inherited money. For these reasons, it is unclear from this study alone whether the association between genetics and wealth inequality is really telling us anything about the importance of genes. Their analysis could be just picking up on population stratification, i.e., on biologically unimportant differences between people from different social classes.

In order to address this question, another study, led by Dan Belsky, a professor at Columbia University, looked at differences between siblings.[28] Belsky and his colleagues were specifically interested in social mobility, defined as the extent to which people had more or less education, occupational prestige, and money *than their parents*. Their study looked at five data sets from around the world, one of which included nearly 2,000 pairs of siblings. They found that the sibling who had the higher polygenic index, who "won" the genetic lottery in the sense that he or she inherited more education-associated genetic variants than his or her co-sibling, was also wealthier at retirement.

These results suggest that if people are born with different genes, if the genetic Powerball lands on a different polygenic combination, then they differ not just in their height but also in their wealth. As was said about Shawn Bradley, some people "won the jump ball of genetic luck"—and winning pays.

Where We Go from Here

Results like these raise a host of scientific questions: How are polygenic indices constructed? Why were these studies conducted using only White people in the United States or using samples of Northern

Europeans? What implications do these results have for understanding racial wealth gaps, which are staggeringly large? (Short answer: none.) Can we really say that genes *cause* you to be wealthier? (Short answer: yes.)

Results like these also raise a host of moral and political questions: Does this mean that differences in wealth are innate or inevitable? That social and economic policies designed to increase equality or redistribute wealth are doomed to failure? This was how the first twin studies on income were interpreted. In 1977, the psychologist Hans Eysenck told *The Times* of London that results on the heritability of income were a sign that governmental agencies tasked with redistributing wealth "might as well pack up."[29] This is, for reasons that I will describe at length in this book, incorrect. But for nearly as long, others have been insisting that genetic research on outcomes like income and wealth is "utterly irrelevant" for social policy. If that's true, why does the link between genes and wealth bother us so much?

In the coming pages, I will attempt to address the litany of questions that I rattled off in the previous two paragraphs. Like any program of research, research on how the genetic lottery shapes our lives has flaws and holes. It makes simplifying assumptions that can't possibly be true; it grapples with incomplete data. Yet this program of research still demands to be taken seriously. As the statistician George Box said, "All models are wrong, but some are useful."[30] There is no shrugging off the fact that children who inherit different genes fare differently in life, in ways that we can measure in dollars and cents. In the next chapter, then, we will dive into how a polygenic index, such as the one used to in the Barth, Papageorge, and Thom study of wealth, is constructed.

Cookbooks and College

When my son was an infant, our pediatrician referred us to a neurologist because she suspected (wrongly, thank heavens) that he had neurofibromatosis type 1. Neurofibromatosis is a rare genetic disorder that causes innumerable tumors in the brain, spinal cord, and at the ends of nerves.[1] Neurofibromatosis is caused by mutations in *NF1*, a gene that codes for a protein that ordinarily prevents cells from overgrowing like tangled strands of ivy. One of the cardinal signs of neurofibromatosis is the presence of café au lait spots—skin blotches that are indeed the color of coffee with milk. My son had two spots.

The suggestion that my new baby might have a serious genetic disease was, of course, terrifying. My then-husband spent the afternoon inspecting our son's skin, over and over again, obsessively checking to see if there were more café au lait spots that we had missed. I was eerily calm all day, and then fell into vivid nightmares when I went to bed. In one nightmare, a silent, angel-like creature gave me pair of tweezers and told me that I could cure my baby of neurofibromatosis. But to do so, I had to pluck out the mutation in the *NF1* gene from each cell in his body, one by one, before morning. I set to my impossible dream task, knowing there was no way for me to get to every cell in time, yet nonetheless frantically poking his tiny body with my magical tweezers.

It was a horrible nightmare. And it captured the essence of what you might think about when you think about genes—the idea that genes have inescapable power, the idea that a person's entire fate can be decided by a single warped molecule. If you have a certain mutation in *NF1*, you will inevitably develop neurofibromatosis. Like Atropos, the Greek Fate who cut the cloth of life with her "abhorred shears,"[2] *NF1* mutations render a certain medical fate irreversible.

In 2013, the journal *Science* published the results of a study that included over 120,000 people and found three genetic variants that were associated with educational attainment, i.e., how many years of school a person completed.[3] Following scientific custom, the three variants have names that seem borrowed from a Stasi filing system: rs9320913, rs11584700, rs4851266.

That result—indeed, the study itself—might seem just as nightmarish as being given a set of a magical, gene-editing tweezers. We have, on the whole, mostly accepted the idea that our genes determine our eye color or rare diseases like neurofibromatosis. But education is achievement. Education is *merit*. And, as I described in chapter 1, education is related to nearly every other inequality in life outcome. The idea that education was fated, that a genetic Atropos named rs11584700 could shear away your achievement, could strip you of your merit—could such an idea be true?

I will argue in this chapter, and indeed throughout this book, that it is a mistake to think of the relationship between rs11584700 and educational attainment as being like the relationship between *NF1* and neurofibromatosis. One's genes do not determine one's educational or financial fate. At the same time, however, it would be a mistake to dismiss the relationship between genes and education as trivial or unimportant. As we saw in the last chapter, one's genetics might not *determine* your life outcomes, but they are still associated, among other things, with being hundreds of thousands of dollars wealthier at the end of one's working life. In order to understand how to take genetic associations *seriously*, without misinterpreting them, we need to understand in more detail exactly what these types of studies are doing (and what they are not doing). So, let's dig into this particular study. How is it, exactly, that researchers concluded that rs9320913, rs11584700, rs4851266 were genetic variants associated with years of education, and how can we interpret that result?

Genetic Recipes, Genomic Cookbooks

One of the lead scientists on the 2013 *Science* paper was an economist, Philipp Koellinger. A tall man, gangly in his forties, with a ready and generous laugh that defies stereotypes about both Germans and economists, Koellinger insists that the group he co-founded, the Social Science Genetic Association Consortium, didn't *really* expect to find anything in their 2013 paper. Rather, they were annoyed at a spate of previous psychology studies that had claimed to find genes associated with intelligence test scores, but had studied so few people that it was statistically impossible that their results were true.

(If you find it surprising that someone would devote years of their life to a project assembling a data set with 120,000 people in order to prove definitively that other people were wrong, then you haven't spent that much time with economists.)

Koellinger's favorite meal is lemon chicken—a whole chicken, nestled in a bed of onions and potatoes, roasted with equal parts lemon juice and olive oil. He made lemon chicken for me the first time he visited Texas and was surprised at how differently it turned out, compared to when he makes it at home in the Netherlands. The giant sour lemons in the H-E-B supermarket are from California instead of Spain. The potatoes here aren't clearly labeled soft-baking versus hard-baking, so he accidentally ended up with potatoes that disintegrated into a delicious, baby-friendly mush. Setting the temperature inside my tiny, ancient oven was an approximate process, at best. And, everything is bigger in Texas, even poultry. Our meal, eaten outside on a warm March evening, was delicious, but a different experience than we would have had if we had followed his recipe for lemon chicken in his Amsterdam kitchen.

All metaphors for genes are wrong, but the recipe metaphor is useful: a gene is a recipe for a protein. Some genes are coding genes, which means they give direct instructions for making proteins. Other segments of DNA act more like the annotations that are penciled in around the recipe. These are the instructions for the instructions, reminding you to, say, get the butter out of the fridge so that it will come to room temperature.

As any home cook can attest, the exact same recipe can yield different dishes, depending on the raw ingredients available and the vagaries of the environment. Roast chicken in Austin is not the same as in Amsterdam. So, too, can the expression of gene into protein differ, from tissue to tissue, from person to person, from environment to environment. Perhaps most importantly, having a recipe in your kitchen drawer won't feed you—something has to *happen* in order to create a final product.

Nevertheless, recipes do constrain your final product. Beginning with a recipe for lemon chicken will not yield, say, chocolate chip cookies. Errors in a recipe can result in a slightly less appetizing dish (not enough salt) or in total disaster (a cup of salt instead of a cup of sugar). In the same way, mutations in DNA sequence can result in slightly altered proteins or in entirely non-functional ones. And, some recipes are more tolerant of error, deviation, and substitution than others. Just as making spaghetti Bolognese does not require the same exacting attention to weight and temperature and timing as making chocolate soufflé, some genes are more intolerant to mutation than others.

All metaphors have flaws, but a recipe is a workable metaphor for understanding the relationship between a single gene and a single protein. For example, *LRRN2* is a gene recipe for the leucine-rich repeat neuronal protein, a molecule that helps keep cells stuck together. But this isn't a book about proteins. This is a book about people. A small change in the sequence of *LRRN2* is associated (very, very, very slightly) with graduating from college, but *LRRN2* isn't a recipe for going to college. We need to stretch the recipe metaphor in a new direction, in order to suggest something about the collective action of lots of genes, in order to build a new intuition about how genes are related to outcomes far removed from molecular biology.

If a gene is a recipe, then your genome—all the DNA contained in the twenty-three pairs of chromosomes in all of your cells—is a large collection of recipes, an enormous cookbook. As I write this, I'm looking at one cookbook on my shelf, *Plenty*, a collection of vegetarian recipes from celebrity chef Yotam Ottolenghi's eponymous London restaurants. Let's suppose you're having a small lunch party with friends at an Ottolenghi café. Your friend just got promoted, or maybe she's announcing that she's finally pregnant. You're digging into platters of roasted eggplant with pomegranate seeds and caramelized fennel with goat cheese. The service is attentive. The company is cheerful and lively. The *Plenty* cookbook is a set of recipes for the food you are putting in your mouth, but it is not, in any straightforward interpretation of the word "recipe," a set of recipes for the party you're enjoying.

Why not? Most obviously, heaps of factors will shape your little lunch party other than the food, from the physical environment (is the lighting too bright or the acoustics too loud or the chairs too back-breaking?) to the social environment (is your friend moody or cheerful?) to culture (is roasted eggplant a familiar comfort food that reminds you of home or an exciting novelty?). The totality of the experience is jointly determined by so many interacting dimensions that some questions are utterly nonsensical. "Which was more important for your dining satisfaction today: including salt in your food or having a chair to sit in?" is a silly question.

In the same way, human lives are jointly determined by interactions between genes and the environment. The classic "nature-nurture debate" is about which one of these is more important. But if we remember that genes are like a recipe that instructs the cook to put salt in your food, and the environment is like having a chair to sit in, we can see that the so-called nature-nurture "debate" is also asking a silly question: genes and environments are always both important.

At the same time, even as we always keep the importance of the environment in mind, we can see that *differences* in genomes are relevant for understanding the differences between people, just as *differences* in cookbooks are relevant to understanding differences between restaurants. The experience of eating at an Ottolenghi cafe is undeniably shaped by which collection of recipes is being used. If the restaurant suddenly switched to serving dishes only from, say, Anthony Bourdain's *Les Halles Cookbook*, your lunch party would be different. This is the basic and incontrovertible lesson we've learned from decades of research on human behavioral genetics. If my genome were different, my cognitive abilities, personality, education, mental health, social relationships—*my life*—would be different, too.

Not all differences are equally consequential. Some differences between genomes are akin to replacing the word "salt" with the word "sugar" throughout a cookbook, or doubling or tripling the amount of salt in every recipe. Huntington's Disease is a good example of this. The *HTT* gene is a recipe for the huntingtin protein, and it contains a segment in which the same bit of DNA is repeated several times. ("Add 1/4 teaspoon cumin. Add 1/4 teaspoon cumin. Add 1/4 teaspoon cumin.") The version of the *HTT* gene that causes Huntington's disease repeats this section of the recipe too many times, resulting in unusually long huntingtin. This protein is then snipped up into small sticky pieces that glob together inside a person's neurons. All of the ghastly symptoms of Huntington's Disease—the depression and anger, the jerking and twitching movements, the eventual loss of a person's basic abilities to walk, talk, and eat—can be traced to this change in the recipe for just one protein.

Most genetic differences between humans, though, are not like replacing salt with sugar or drastically changing the quantity of a vital ingredient. They are like replacing the word "onion" with the word "leek." The challenge for science is to understand whether such tiny changes in the genome are actually meaningful for understanding the differences between human lives, and if so, why.

One Ingredient at a Time

Let's leave the question about human lives aside, for now, and instead follow our restaurant metaphor a little bit further. One way you could try to figure out whether tiny recipe changes make a difference to people's restaurant experiences would be begin with what you already know about cooking and eating, and then narrow in on an ingredient that you think might be important. Something like . . . cilantro? Some people think cilantro tastes like soap and will carefully scrape each tiny leaf off their tacos. Armed with the knowledge that some people don't like cilantro, you could then pick twenty restaurants in town and measure whether or not there is cilantro in any of the dishes they serve.

Yes, it's only twenty restaurants, which really isn't that many, but to compensate for how few restaurants you've included in your study, you can measure people's restaurant experience very carefully. You don't just ask people how much they liked their meal. You send a trained investigator to count how many times people smile and laugh while they are eating; you look at their credit card bills to see how much money they spend on that restaurant in a year. At the end of the day, you have data on just one ingredient, measured in just a few restaurants, but you understand the outcome you're trying to measure really, really well. You can then proceed to test whether your hypothesis was right. Do restaurants that don't serve cilantro have happier customers?

This strategy—beginning with some *a priori* knowledge about biology, zeroing in on one genetic ingredient, and measuring an outcome very intensely in a relatively small number of people—is the strategy that many psychologists and geneticists took in the early 2000s. This approach was called a *candidate gene* study. Perhaps the most famous candidate gene variant is called 5HTTLPR, or the serotonin (which is abbreviated, confusingly, as 5HT) transporter-linked polymorphic region.[4] The idea was relatively simple. We already think that serotonin is involved in depression because if you give people antidepressants that target serotonin (like Prozac), they get less depressed (sometimes). 5HTTLPR is a tiny part of the genome that affects how serotonin is shuttled between neurons in your brain, so perhaps people who have different versions of the gene will differ in their vulnerability to depression.

To take things one step further, we also know that people who are stressed out are more likely to get depressed. Divorce, job loss, poverty, an abusive childhood—stressors like these are the most robust predictors of depression. So, perhaps a 5HTTLPR variant doesn't cause you to become depressed *unless* you are already stressed in some way.

Tens of millions of dollars were spent chasing that hypothesis. Researchers poured money not just into measuring people's DNA, which was—and is—getting cheaper every day, but into carefully measuring everything about people's brains and minds. Whether they met a doctor's definition of being clinically depressed, yes, but also how quickly they recalled sad memories and how many milliseconds their eyes spent looking at sad pictures and what parts of their brains lit up when they were listening to sad music. Across hundreds, even thousands, of scientific studies, results kept coming, about how a 5HTTLPR variant caused stressed people to get depressed.

The problem was that all of those results were wrong. By 2019, after years of polite warnings, statistical grumblings, and escalating annoyance at the seemingly-endless fount of 5HTTLPR studies, the psychologist Matt Keller published a study, the title of which did not beat around the bush: "No support for historical candidate gene or candidate gene-by-interaction hypotheses for major depression across multiple large samples."[5] Liberated from the polite conventions of scientific journals, the psychiatrist and blogger Scott Siskind summarized the paper's conclusion more colorfully. He denounced investigators who reported "results" on 5HTTLPR as fabulists telling stories about unicorns, except worse: "This isn't just an explorer coming back from the Orient and claiming there are unicorns there. It's the explorer describing the life cycle of unicorns, what unicorns eat, all the different subspecies of unicorn, which cuts of unicorn meat are tastiest, and a blow-by-blow account of a wrestling match between unicorns and Bigfoot."[6]

The major problem with the candidate gene approach, which seems obvious in retrospect, is that there is no one gene *for* depression, any more than there is one ingredient for having a successful restaurant. There are not even ten genes for depression. Depression, body size, graduating from college, impulsivity, even height—these are all *complex* traits, meaning that, in one critical way, they aren't like Huntington's disease at all. They aren't caused by a single gene. They are influenced by thousands upon thousands of genetic variants, each of which has a minuscule effect. And because the effects are so minuscule, you need to study many, many, many more people than were included in any of the early candidate gene studies.

An early, high-profile study of 5HTTLPR published in 2003 included 847 people; the definitive rebuttal of that study published in 2019 included 443,264—about *500 times* more people.[7] The first study to report any genes that were reliably associated with depression (using a method that I'll describe later in this chapter) included 480,359 people.[8] A cruel consequence of trying to find tiny patterns when you don't have nearly enough data (when, in fact, you have less than 1 percent of the data you really need) is not just that you risk missing the patterns that are really there—you also risk picking up on "patterns" that seem real but are actually just noise.

The Cookbook-Wide-Association Study

OK, so the candidate gene thing didn't work. But you still want to figure out which—if any—genetic variants make a difference for people's lives. You want to identify tiny recipe changes that make a difference to the restaurant experience. Plan A—beginning with your knowledge of cooking and coming up with plausible-seeming hypotheses and working from there—initially seemed clever but turned out to yield absolutely no valuable knowledge. As a testament to how bad scientists can be at forecasting the future, Plan B—declaring your previous knowledge of cooking to be useless and abandoning all attempts at coming up with plausible-seeming hypotheses—initially seemed not so much clever as entirely ludicrous. Yet it was this approach that finally began to yield results.

Instead of beginning with a single ingredient measured in a few restaurants, let's instead imagine that you took every recipe for every dish served at every restaurant in Austin, Texas, and broke it down into tiny elements. For each restaurant, the resulting dataset would be huge. There would be thousands, maybe hundreds of thousands of rows, representing quantities, ingredients, times, temperatures, instruments, and instructions. Chop finely. Bake at 300 degrees. 1 tablespoon. Cumin. Sauté. Golden brown.

And—like the genome—most of that data would be *exactly the same* across restaurants. Human DNA is more than 99 percent the same across all humans. Every restaurant uses salt.

So, let's filter out all the recipe elements that are exactly the same across restaurants. Keep only the things that differ between them. A restaurant near my house serves a fried bologna sandwich with giardiniera, the Italian relish of pickled vegetables. You're not going to find giardiniera everywhere.

Now that we've broken down the food being served at each restaurant into bite-size bits of data, we need a measure of the restaurant itself. But we no longer have a boutique sample of twenty restaurants, so we don't have the time or money to do boutique measurements of customer satisfaction. We need something quick and dirty, something easily compiled across thousands of restaurants. Perhaps something like . . . Yelp ratings?

In case you haven't been on the internet in the past decade, Yelp is a website that crowdsources information about local businesses. People leave reviews and rate the business from one to five stars. For a restaurant to be highly rated on Yelp, lots of people need to have enjoyed eating there enough to leave a comment and a good review, and the restaurant can't have too many miserably dissatisfied customers. As of this writing, the #1 Yelp-rated restaurant in Austin, Texas, is a gastropub named Salty Sow, which, as its name suggests, is a temple to all things pork—candied pork belly, collard greens with ham hocks, deviled eggs with bacon. #2 is a Southern cafe known for its chicken-fried steak, and #3 is Franklin Barbecue, where people queue for hours for slices of beef brisket with perfectly charred bark.

Now, all we need to do is take the recipe-elements data we compiled in the first step, combine it with the data on restaurant Yelp ratings, and voilà! We have the makings of a statistical analysis. Are restaurants that serve giardiniera rated higher?

But wait! A chorus of objections rises to a fever pitch. Candied pork belly? Chicken-fried steak? Beef brisket? The vegan reminds us that Yelp ratings don't reflect certain values that we maybe should embrace about food, like ethical objections to eating meat. The bespectacled hipster grumbles that prioritizing mass market satisfaction is antithetical to creativity. And the sociologist reminds us that people who spend lots of time posting anonymously on the Internet aren't *really* a representative sample of customers.

So, we consider other alternatives. Instead of Yelp rating, perhaps revenue is a better measure? The Perfect 10 Men's Club rakes in a lot of money relative to other Austin businesses that serve food and alcohol, but I suspect customers are coming for something other than the food. Maybe expert opinion is a superior measurement of restaurant quality? The Top 20 list by *Condé Nast Traveler* does, after all, include several of my personal favorites. But then again, perhaps the food preferences of middle-aged professors aren't the most widely generalizable.

The chorus of objections is right. Measurement *does* matter. Which cookbook words are correlated with being the "best" restaurant will depend on how you define "best." Expert opinion might pick up on exotic ingredients (grated sea urchin, anyone?), whereas profits might be picking up on food that can be cheaply and reliably prepared by workers with little prior skill or training (or the fact that the food is served while customers watch scantily clad dancers).

In psychology, we spend a lot of time thinking about these sorts of measurement problems. You have a theoretical entity, or *construct*, that you are interested in studying (e.g., how much do people enjoy a restaurant?). And you need to come up with a way of meaningfully attaching a number to that construct. What measurement is the best? Some measurement problems are simple: Height is measured in inches. But often the constructs we are interested in are ineffable and controversial. What is the unit for happiness? What does it mean to be smart? What makes a good restaurant?

Some of you might recoil at the idea of trying to quantify restaurant quality at all, no matter how you do it. Food is endlessly varying. How can one hope to reduce the tapestry of sensorial and cultural experience that is a city's restaurant scene into a single number? In the context of genetics research, people often object to the measurement of intelligence or personality on similar grounds: people are also endlessly varying, and how can you reduce all of their quirks and talents into a single number?

The short answer is that you can't. Fortunately, "reducing" entities, whether they be people or restaurants, is *not* the goal of measurement. Measurement is the process of assigning numbers to events or characteristics, and it is essential to *all* of science. You cannot study anything scientifically unless you can measure it. A restaurant's Yelp rating is not the end-all, be-all measure of its value, and the idiosyncrasies of what makes a restaurant someone's favorite local joint might not be reflected in how many stars it's been given by anonymous people on the Internet. But even if we wholeheartedly reject the idea that everything valuable or interesting about a restaurant can be reduced to its Yelp rating, we can consider it a crude, but nonetheless useful, metric for how much people generally say they enjoy eating at a particular establishment.

For it to be useful, however, we need to be clear-eyed about its flaws and limitations. A study that measures restaurants using Yelp reviews will give you an imperfect measure of how many people say they liked eating at a restaurant, but the resulting rank-ordered list of "good" restaurants might not reflect a person's values about what constitutes "good" food.

When I teach Introduction to Psychology, I have students practice talking about psychology studies using the following language: "This study was about Construct X, as measured by Y." For example, this study was about happiness, as measured by people's ratings on an item that asked how happy they felt today. Or, this study was about social anxiety, as measured by how much cortisol in saliva increased when people were asked to do a short speech in front of unsmiling judges. This language exercise, hopefully, helps them learn to pay attention to how researchers measure abstract concepts like happiness and anxiety, and to be curious about how those measurements might be flawed.

At the end of the day, however, in order to study something scientifically you have to measure it, and those measurements are constrained by the practicalities of time and money. So, let's return to our cookbook-wide association study of restaurant quality, as measured by average Yelp rating. You run millions of correlations: Which of the recipe elements that *differ* between restaurants are correlated with a restaurant's Yelp rating?

You would get, as a set of results, something that looked like a deranged shopping list. And each entry in the list would be accompanied by a small number that represents how strongly it is correlated with a restaurant's Yelp rating. Huge recipe errors with really big effects, like substituting sugar for salt, are going to be rare to the point of being practically nonexistent. If a restaurant served food with no salt, it probably wouldn't stay in business long enough to be in your database.

What this analysis will pick up on, if anything, is tiny but consistent patterns. Maybe these patterns will confirm some of your previous hunches. Maybe, as the chef Mario Batali claimed, "crispy" is the word that sells the most food. Or, maybe the results will reveal patterns that no one had ever thought of before. Surprising or not, the resulting correlations will be very small. The differences between Whataburger and Shake Shack probably cannot be reduced to a single word in one recipe.

Regardless of how our cookbook-wide-association-study turns out, however, a few things will be obvious. The results *don't* imply that the restaurant's environment—its seating and music and lighting and location and decor—are unimportant to people's experiences. The analysis won't tell you whether a website where strangers write anonymous reviews of local businesses is a *good* thing, or whether the ratings are fair. It won't tell you whether those ratings accord with your ethical and aesthetic values, or how a restaurant *should* be run. And the results will definitely not teach you how to cook. What it will teach you, very simply, is what recipe elements are more common in the restaurants that you have classified as being "high" versus "low" on some metric.

From CWAS to GWAS

The cookbook-wide analysis that I've just described is basically how a genome-wide association study (GWAS, pronounced "JEE-Wos") works. Just like an analysis that correlates individual cookbook words with some measurable characteristic of restaurants, a GWAS correlates individual elements of the genome with some measurable characteristic of people. The individual elements of the genome cookbook that are most commonly analyzed are called single nucleotide polymorphisms, abbreviated SNPs (pronounced "snips").

A DNA molecule is made up of two sugary strands, zipped together with interlocking pairs of four different types of nucleotides— guanine (G), cytosine (C), adenine (A), and thymine (T). A SNP is a genetic difference between people where some people have one nucleotide at a particular spot (locus) in their genome, whereas other people have a different one. You might have a G whereas I have a T. The different versions of the SNP are called *alleles*. Typically, one allele is more common than the other; the allele that is less common in the population is the *minor* allele. Everyone has two copies of every gene (one from their mother, one from their father), so you can count the number of minor alleles that a person has for each SNP (0, 1, or 2). A GWAS measures millions of SNPs in thousands of people and correlates each SNP with a *phenotype*, i.e., with something that you can measure about a person, like height, body mass index, or years of education.

Even millions of SNPs are only a fraction of the total pool of genetic variation that exists between people. But a GWAS can typically get away with analyzing data from only a fraction of the genome, because each SNP that is measured "tags" many other genetic variants that differ between people. When you read a recipe that includes the words "black pepper" you can make a decent guess (you can "impute") that the recipe probably includes the word "salt." Similarly, when people have one form of a SNP, you can often reasonably impute information about other, nearby genetic variants: If someone has a "C" in a particular location, you can often reasonably impute that they have a "T" in another location. (Rare variants, which only occur in a few people or families, will be largely untagged.)

This ability to tag multiple variants by measuring just one is the result of how egg and sperm cells are made. Recall the difference between humans and bacteria: rather than simply copying ourselves, DNA letter-for-letter, we reproduce sexually. That is, we create sperm or egg cells, each of which contains *half* of our DNA, with the other half of the genome necessary to create a whole person coming from the other parent. But in making sperm or egg cells, our bodies do not package one intact copy of an entire chromosome. Rather, in the process of meiosis, something called *recombination* occurs. For each of my twenty-three pairs of chromosomes, the chromosome that I inherited from my mother and the chromosome that I inherited from my father line up and trade pieces. This recombination process does, in fact, *re-combine* genetic variants into brand new combinations that are all different from each other.

The recombination process is the biological basis of what Gregor Mendel, on the basis of his mathematical observations, called the law of independent assortment. The probability of inheriting a certain version of gene A is independent from the probability of inheriting a certain version of gene B.

Except, that is, when gene A and gene B are very close to each other, physically, on the genome. Recombination shuffles the metaphorical "deck" of paternal and maternal chromosomes, but it does so badly, leaving chunks of cards stuck together. When genes are physically closer, the chances that recombination will occur somewhere in the space between them are smaller. Very physically close genes are likely to be inherited together rather than separated by the recombination shuffle. Genes that are likely to be inherited together, by virtue of their physical proximity to one another, are said to be in *linkage*. Linkage leads to genes being correlated with one another, i.e., being in *linkage disequilibrium*, or *LD*.

Understanding the LD structure of the human genome has turned out to be extraordinarily useful for many purposes, including making the measurement of the genome more efficient in GWAS. But if a GWAS discovers that a particular SNP is associated with, say, going further in school, it is unclear whether that association is driven by that particular SNP itself, or whether the signal that one is detecting is due to the fact that the SNP you measured is co-inherited with a variant that you *didn't* measure.

One way to think about GWAS results, then, is that they are a genetic treasure map: X marks the spot, and now you know that the pirate gold is located in the southwest part of a particular desert island. Once you land on that island, though, you find a jungle of vines and densely packed trees, and there's no getting around labori-ously trawling over every inch of the jungle, a.k.a. "fine-mapping" the island.

The 2013 study that I told you about at the beginning of this chap-ter, which found three genetic variants associated with whether or not a person graduated from college, was a GWAS. And if you think of a GWAS as being an incredibly crude way of reading people's genomic cookbooks, it's easier to make intuitive sense of the result. There are important parallels between genetic research on educa-tion and our hypothetical cookbook-wide-association study that are worth considering.

First, educational attainment, defined as the number of years of formal schooling that a person completes, is rather like a restaurant Yelp rating. On the one hand, both have real consequences. Restau-rants with low Yelp ratings are less likely to stay in business. People without diplomas are less likely to be employed. On the other hand, neither metric fully represents all the characteristics that we might cherish and value. High scorers could even have characteristics that we find deeply objectionable. A restaurant might have a good Yelp rating because it serves good-tasting food in a pleasant atmosphere. But it might also be a national chain that serves factory-farmed meat and caters to tourists, pushing out local establishments that have more charm. A person might go far in education because they are smart and curious and hard-working, or because they are conform-ing and risk-averse and obsessive, or because they have features (pretty, tall, skinny, light-colored) that privilege them in an intrac-tably biased society. A study of what is correlated with succeeding in an education system doesn't tell you whether that system is good, or fair, or just.

Second, *who* rises to the top of an educational system is cultur-ally and historically specific. A cookbook-wide-association might not find the same correlates of highly rated restaurants in Austin as it would in Manhattan (not to mention in Delhi or Shanghai). In the same way, a GWAS of educational attainment in, say, American men who were born after 1970 will detect genetic variants that are associated with more education *in that population*. Whether these same genetic variants are also correlated with more education in, for example, American women who came of age before sex discrimination in higher education was outlawed, is an open question. This inconsistency in findings across contexts doesn't mean that Yelp ratings are a terminally flawed measure of how much people like restaurants, or that GWAS results are useless. It does, however, mean that results you obtained using a measure taken in one time and place don't necessarily generalize everywhere.

Third, GWAS results do not, *in any way*, show that environments don't matter to education. GWAS doesn't even measure anything about the environment.

Fourth, GWAS, on its own, will not show that we understand education "at the level of DNA," any more than an automated analysis of cookbooks would prove that we understand socializing in restaurants "at the level of the ingredient."

Fifth, the correlations that a GWAS detects are very, very, very, *very* small. As well they should be: The differences between Whataburger and Shake Shack cannot be reduced to a single word in one recipe, and the differences between someone who finishes a PhD and someone who drops out of high school cannot be reduced to a single SNP. Each SNP associated with educational attainment is worth, at most, only a few extra weeks of schooling, and most SNPs are worth much less.

Sixth, GWAS results will not tell you how to cook. That is, the list of genomic ingredients (SNPs) does not tell you the mechanisms for how those ingredients come together to make a complicated outcome.

Nightmarish or Negligible?

When you consider how marginal associations with a single SNP are, along with the fact that one cannot readily tell whether a "significant" association is even driven by the measured SNP itself or another variant that is "tagged" by that SNP, a genetic study of educational attainment stops seeming nightmarish. It instead starts feeling as trivial as a study correlating recipe words with restaurant Yelp ratings. Which is to say, pretty trivial. SNP rs11584700 isn't your genetic fate. It can't even get you past the first midterm of freshman year. Which leads many people to wonder: Well, why bother doing GWAS at all? If you focus on the magnitude of per-SNP effects on education, it's easy to dismiss the whole GWAS enterprise as a waste of time and money.

This is some people's take on GWAS—not that it's scary or eugenic, but that it's trivial and wasteful. My PhD advisor, Eric Turkheimer, for instance, is famous for his skepticism about the value of GWAS. In 2013, as outgoing President of the Behavior Genetics Association, he gave a now infamous speech at the association's annual conference dinner. The Marseille sky was pink and gold; people were wearing their suit jackets and party dresses. The mood was convivial—until Eric got on stage and said that conducting GWAS was like studying the pits on a music CD to try to understand whether a song was any good. Given to an audience of fellow scientists who had collectively invested millions of dollars and years of their life to GWAS research, Eric's speech did not go over well.

I wasn't shocked, because I had heard a version of his argument many times before. And as of 2013, I probably would have (secretly) agreed with him. Year after year, I had listened to scientific talk after scientific talk about efforts to find genes associated with interesting life outcomes. All of these ended pretty much the same way—"We haven't found anything *yet,* but just wait until we get more people!"

The string of what seemed like expensive failures, combined with, to be honest, my resentment at how genomic methods had seemed to usurp the traditional tools of behavioral genetics (twin and family studies), which I had just spent a decade mastering, made it easy to agree with him. Human behavior was just *too complex*. There's no way we are going to learn about Debussy from studying the pits of a CD. There's no way we can learn anything useful about what makes a great restaurant great by doing a cookbook-wide-association-study. And there was no way we were going to learn anything useful about real human lives from correlating them with SNPs.

But we were wrong.

Polygenic Indices and the (Un)predictability of Life Outcomes

Part of the secret to understanding what makes GWAS valuable is to go back to something I've already told you: that the association between any individual SNP and educational attainment is very, very, very, *very* small. This result might seem obvious—inevitable, even. But it's a conclusion that defied the predictions of thousands of scientists around the globe. In the early 2000s, when the GWAS methodology was first being developed, many scientists predicted that phenomena like schizophrenia or autism would be caused by perhaps a dozen different genetic variants.

If that were true, then the genome would readily give up its secrets in studies of just a few thousand people, and the effect of each gene would be relatively large. But those early predictions turned out to be comically naïve. Schizophrenia and autism and depression and obesity and educational attainment are not associated with one gene. They are not associated with even a dozen different SNPs. They are *polygenic*—associated with thousands upon thousands of SNPs scattered all throughout a person's genome.

Maybe the clearest example of massive polygenicity is height. As we discussed in the last chapter, very tall people, like the basketball player Shawn Bradley, can be tall because they inherited very many height-increasing genetic variants. Height might sound like a somewhat boring trait, but it is studied a lot in statistical genetics. Height is easy to measure accurately. It's highly heritable. Nearly every biomedical study collects data on height, so researchers can curate huge samples of people. And the lack of controversy regarding the fact that genes influence how tall you are allows scientists a bit of space to work out their mathematical models.

One such mathematical model calculated that over 100,000 SNPs might each have a small association with how tall you are.[9] On the basis of those calculations, the authors proposed what they called the "omnigenic" model—*omni-*, of course, meaning *all*. All the genes. Or rather, all the genes whose recipes are read in the bodily tissues that are involved in whatever it is you're studying. If you're studying height, the relevant bodily tissues include the pituitary gland (which produces growth hormones) and the skeletal system, among other things. If you're studying education, the relevant bodily tissue is the brain.[10]

As outcomes become more polygenic, the number of people you need to study, in order to differentiate signal from noise, increases accordingly. If even a trait as ostensibly simple as height is polygenic, then we should expect that social and behavioral outcomes are equally polygenic—if not even more so. Taking the implications of polygenicity seriously, the first study of educational attainment, which I told you about at the beginning of this chapter, included what at the time seemed like a shockingly large number of people—126,559—and discovered 3 SNPs that were associated with years of education. The same research group persisted, and their second study, published just three years later in 2016, included 293,723 people and found 74 genome-wide significant SNPs. And the third follow-up study, published in 2018, included 1.1 million people and found 1,271 genome-wide significant SNPs.[11] The prediction that GWAS could, in fact, find SNPs that were reliably associated with a complicated outcome like educational attainment, *if* the study simply had enough people, turned out to be true.

When thousands of genetic variants are involved in a trait, tiny correlations with each one can add up to meaningful differences between people. And that is exactly what researchers do: add up the information across all SNPs into a single number. More specifically, after you've conducted a GWAS, you have a long list of numbers, one for each SNP that you've measured, that represents the strength of the relationship between each SNP and your target phenotype (in this case, educational attainment). You can then take that list of numbers and use it as a type of scoring key for DNA from a new group of people. How many copies of each education-associated genetic variant does each person have—0, 1, or 2? (Remember that you get two copies of every gene, one from your mother and one from your father.) The number of copies of each SNP is multiplied by the strength of its relationship to educational attainment, and then

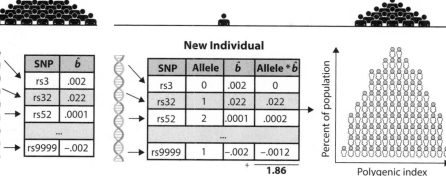

Discovery GWAS

N = 30K–1,000K

Polygenic scoring

N = 300+

New Individual

SNP	$\hat{b}$
rs3	.002
rs32	.022
rs52	.0001
...	
rs9999	–.002

SNP	Allele	$\hat{b}$	Allele * $\hat{b}$
rs3	0	.002	0
rs32	1	.022	.022
rs52	2	.0001	.0002
...			
rs9999	1	–.002	–.0012

+ **1.86**

Percent of population

Polygenic index

FIGURE 3.1. Creating a polygenic index. Figure reproduced from Daniel W. Belsky and K. Paige Harden, "Phenotypic Annotation: Using Polygenic Scores to Translate Discoveries from Genome-Wide Association Studies from the Top Down," *Current Directions in Psychological Science* 28, no. 1 (February 2019): 82–90, https://doi.org/10.1177/0963721418807729. Correlations between individual SNPs and a phenotype are estimated in a "Discovery GWAS" with a large sample size. Many GWAS have samples that exceed millions of people. Then, a new person's DNA is measured. The number of minor alleles (0, 1, or 2) in this individual's genome is counted for each SNP, and this number is weighted by the GWAS estimate of the correlation between the SNP and the phenotype, yielding a polygenic index. This polygenic index will be normally distributed: most people will have an average polygenic index, but a few people will have very low or very high scores. Reprinted by permission of SAGE Publications, Inc.

everything is added up across all the SNPs that have been measured in your genome (figure 3.1). This composite is a *polygenic index.*

The researchers of the educational attainment GWAS mentioned above then computed educational attainment polygenic indices in a new sample of participants. All of them were White Americans who were high school students in the 1990s. Among those who had the lowest polygenic indices, the rate of graduating from college was 11 percent. In contrast, among people with the highest polygenic index, the rate of graduating from college was 55 percent. That kind of gap—a fivefold increase in the rate of college graduation—is anything but trivial.

I've described the strength of the relationship between the polygenic index and educational attainment in terms of the difference in rate of college completion between people who are low on the polygenic index and people who are high. A different way to express the strength of the same relationship—what statisticians call the "effect size"—is in terms of what is called an *R-squared* (R^2), a measure of how much variation between people in one thing can be captured by something else you've measured about them. For example, people differ in their body weight. Taller people weigh more—but how much of variation in people's body weight can be accounted for by knowing their heights?

The R^2 is expressed as a percentage than can vary between 0 and 100. In the example of height and weight, if the R^2 were 100 percent, that would mean that people varied in their body weight only because they were taller or shorter than other people, and that one would know how much a person weighed just from knowing their height. If the R^2 were 0 percent, that would mean that *none* of the differences between people in how heavy they were related to how tall they were, and that knowing someone's height gave no information about their weight. In reality, the R^2 is most commonly in between these extremes. In the US, differences in height account for about 20 percent of the variation in how much people weigh. In other words, about one-fifth of the variation between people in how much they weigh can be captured by knowing their heights—but there are still differences between people who are equally tall.

Researchers talk about R^2 values using language that can be confusing and misleading. Two words, in particular, are troublemakers. One word is "explain." An R^2 is often referred to as a "variance explained," but in my view, "explanation" implies a much deeper understanding of *how* two things are related. An R^2 value isn't doing any scientific explanation; it's simply a mathematical expression of how strongly one variable co-occurs with another variable.

The second troublemaking word is "predict." In the course of ordinary conversation, when we talk about the ability to "predict" the weather, or the outcome of an election, or stock market activity, we are usually implying that those forecasts of future events are highly accurate. In contrast, researchers often use the word "predict" and "predictor" when their forecasts about the future are highly uncertain and frequently inaccurate. In the example of height and weight that I gave you, I can statistically account for variation in how much people weigh using information about how tall they are: Height is a "predictor" of weight. If I'm forced to guess how much someone weighs, my best guess will be better if I know how tall they are than if I don't have that information. But my best guess will still be pretty bad, because people who are the same height can still vary by a lot in their weight.

With this information in mind, what is the R^2 for a polygenic index and educational outcomes? In samples of White people living in high-income countries, a polygenic index created from the educational attainment GWAS typically captures about 10–15 percent of the variance in outcomes like years of schooling, performance on standardized academic tests, or intelligence test scores.[12]

Whether 10–15 percent sounds like a lot or a little depends very much on your perspective. Often, in my experience, there is a rush to minimize an R^2 of 10 percent as trivial and potentially ignorable. Certainly, an R^2 of 10 percent belies any characterization of polygenic indices as "fortune-tellers" that will accurately predict the future for any individual person.[13] In figure 3.2, I've graphed a hypothetical relationship between a polygenic index and a life outcome where the former captures about 10 percent of the variance in the latter. If you pick any point on the horizontal axis, i.e., if you pick any given value of the polygenic index, and then scan up and down, you will see that there is still a lot of variability in people's life outcomes. This matches what is seen in reality: some people with an average polygenic index have a PhD, some didn't graduate from high school, and everything in between.

But while polygenic indices are not perfect "fortune tellers" for individual lives, neither can they be dismissed as trivial or negligible. As the psychologists David Funder and Daniel Ozer argued, we can ground our intuitions about whether an R^2 value is "big" or "small" by benchmarking it against the strength of relationships encountered in everyday life[14]—such as the tendency for antihistamines to relieve allergy symptoms ($R^2 = 1\%$), for men to weigh more than women ($R^2 = 7\%$), for places at higher elevations to be

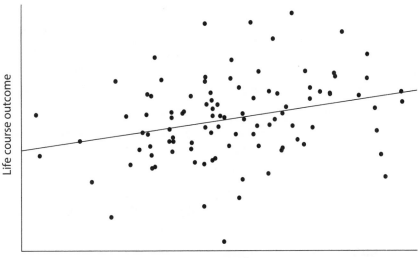

FIGURE 3.2. Hypothetical polygenic index that captures 10% of the variance in a life course outcome. Polygenic index on the horizontal axis; hypothetical life outcome, such as educational attainment, on the vertical axis. Each dot represents an individual person. For each value of the polygenic index, there is considerable variability in people's life outcomes.

colder ($R^2 = 12\%$), and as previously mentioned, the tendency for taller people to weigh more ($R^2 = 19\%$). To this list, we can add a benchmark specifically relevant to the study of social inequality—the tendency for children born to richer families to graduate from college at higher rates ($R^2 = 11\%$).[15]

The comparison with income is particularly poignant, because we are so accustomed to thinking about the ways that money can advantage students. Parents with more money can buy more toys and books, send their children to better schools, sign them up for art classes and robotics after-school programs. They can afford private tutors and SAT prep courses. Students from affluent families don't have to work their way through college, and they have more time to focus on their studies. Certainly, a child who is raised in a rich family isn't *destined* to get a college degree. Your family's financial circumstances don't fully *determine* your social class in adulthood. But money matters for understanding who is most likely to get a college education in the United States and who is least likely.

The numbers from these benchmark comparisons might be smaller than you imagined they would be. In their paper, Funder and Ozer describe three reasons why R^2 values are generally lower than we might anticipate. First, and most simply, humans are very different from one another. There is a lot of variability to explain.

Second, human lives are causally complex, the result of a multiplicity of interacting factors. Given the sheer number of potentially relevant factors, it is simply not realistic to expect any one variable— even something as important as income or, yes, genetics—to account for more than a fraction of the variance in an outcome. In Funder and Ozer's words, "perhaps all researchers should lower their expectations a little (or a lot)."[16]

The need for researchers to lower their expectations regarding *any* variable, environmental or genetic, was underscored by a recent study called the Fragile Families Challenge. The Fragile Families and Child Wellbeing Study is an ongoing study of over 4,000 families who were recruited for a study of child development when their children were born. The children have since been measured on a raft of variables when they were 1, 3, 5, 9, and 15 years old—their parents, teachers, and eventually the children themselves were surveyed about, e.g., "child health and development, father–mother relationships, fatherhood, marriage attitudes, relationship with extended kin, environmental factors and government programs, health and health behavior, demographic characteristics, education and employment, and income; parental supervision and relationship, parental discipline, sibling relationships, routines, school, early delinquency, task completion and behavior, and health and safety."[17] In other words, *everything* about a child's environment and development that researchers could possibly think to measure.

Right before the study investigators released the data from the measurements taken at age 15, they devised a challenge: Teams of scientists were tasked with trying to predict the children's outcomes at age 15, using as many variables and as fancy statistical methods as they wanted. Ultimately, over 160 teams of scientists participated in the challenge, and each team was given access to over 12,000 variables about a child and their family. The results were sobering. The best model—which, again, could potentially incorporate *thousands* of variables measured about a child since their birth—predicted just 20 percent of the variance in students' grades at age 15. Describing these results, the organizers of the Fragile Families Challenge echoed Funder and Ozer's call for humility when studying complex human lives: "If one measures our degree of understanding by our ability to predict, then results . . . suggest that our understanding of child development and the life course is actually quite poor."[18]

Any discussion of whether genetic associations with human life outcomes are "strong" or "weak," then, must grapple with the fact that, when researchers are studying something as complicated as children's school performance, no variable or set of variables has an impressive-looking R^2—not even when researchers measure every aspect of the environment that they can imagine.

Yet, even a small R^2 can be quite meaningful. The third reason that Funder and Ozer give to make sense of why R^2 values are often smaller than we imagine is that small effects can *add up* when they are repeated, time and time again, person after person. We are accustomed to thinking about how the small but systematic effects of income cumulate—at every point in an educational trajectory, wealthy families have their thumb on the scale, making it just a bit more likely their children will experience a certain outcome (a certain grade on a test, placement in a more advanced class, admission to a selective school). Multiplied across millions of families, this process results in an unignorable social pattern of educational inequality. So, too, might DNA be a systematic force in a child's life, putting its thumb on the scale at every point in an educational trajectory. And the advantages bestowed by having a certain combination of DNA variants cumulate across millions of people in a population, similarly resulting in an unignorable pattern of educational inequality. *In the long run*, what look like small effects can be meaningful.

The ability of polygenic scores to "compete" statistically with the other variables that we already think of as important for the study of social inequality—variables like family income—changed my mind about the value of GWAS. Knowing that a SNP named rs11584700 is associated with staying in school for an extra two days might not be particularly valuable. But a polygenic index, where the tilt in educational outcome between the highest-scoring students and the lowest-scoring students is just as steep as the tilt between the richest and the poorest students, *is* valuable. As we will see in the coming pages, this development opens up a new landscape of research possibilities—and unleashes a new avalanche of interpretive questions.

In the chapters to come, we will begin to tackle those interpretive questions one by one. We will consider whether these polygenic associations are causal (chapters 5 and 6), what might be the mechanisms for how genes influence something as complicated as education (chapter 7), and what, if any, implications these results have for our ability to change people's educational trajectories (chapter 8).

Before we dive into those questions, however, you might have noticed that the studies I've been describing have all been conducted using samples of people who all identify as White. The genetic studies are telling you something about how *individuals* differ within a population that is homogeneous with regard to race. At the same time, some of the biggest disparities in educational outcomes or income are seen *between* racial groups. As I already mentioned in the introduction, it would be a grave error to think that the genetic results I've described give us any information about the causes of between-group differences—but why? In the next chapter, we will consider this question.

4

Ancestry and Race

In the lecture on memory in my undergraduate Intro Psych class, I ask my students to remember a list of words. It includes words like "dream" and "bed" and "rest." Then I ask them to write down the words they remember. Invariably, they (mis)remember hearing the word "sleep"— even though I never said the word "sleep." The idea of "sleep" is activated in the brain because other words in the same semantic network, words that have been associated with sleep through relentless repetition, have also been activated. The word "sleep" is retrieved as if it were really *heard*, even though it was never said.

When people hear "bed," they cannot help but hear "sleep." When people hear "genes" or "intelligence"—particularly in the United States—they cannot help but hear "race." A reader new to this topic might therefore be surprised to learn that there is zero evidence that genetics explains racial differences in outcomes like education. Currently, stories about genetically rooted racial differences in the complex human traits relevant for social inequality in modern industrialized economies—traits like persistence and conscientiousness and creativity and abstract reasoning—are just that. They are stories.

Nevertheless, discussing race and racism in a book about genetics is necessary. From a very early point in its scientific development, the study of heredity became braided into racist ideas in order to justify racist actions, and the enthusiastic appropriation of genetics by racists continues well into the twenty-first century.[1] To avoid any mention of race leaves a vacuum that would be filled with errors and that would be interpreted as a tacit approval of scientific racism. At the same time, because discussions of genetics in relationship to class structure and to redistribution of resources have been poisoned by decades of race "science," well-intentioned people often feel that they need to reject information about genetic influences on social and economic outcomes outright, in order to preserve their commitment to antiracism. It is critical, then, to separate the empirical reality of genetic influences on *individual* differences in socioeconomic attainments from the racist rhetoric about differences between human groups.

In this chapter, I aim to clarify why today's genetically inflected incarnation of scientific racism is both empirically wrong and morally blinkered. I will first describe what geneticists mean by *ancestry* and why it is false to collapse the idea of ancestry with race. I will then describe how genetics research has been done *by* predominantly White scientists using predominantly White research participants—a situation that creates conditions for false comparisons between racial groups and risks exacerbating inequities between racial groups. I also describe why it's wrong to assume that research on the genetic causes of *individual differences* within a population gives us information on the causes of *group differences,* a statistical fallacy that is commonly buttressed by racist presumptions about White supremacy. And I conclude by looking ahead at the coming avalanche of multi-ethnic genomic data, and describe why we need not fear that any statistical result will compromise a commitment to antiracism and racial equality.

We're All Descended from Everyone in the World

My grandparents are Pentecostal Christians who have memorized large portions of the Christian Bible and who encouraged their children and grandchildren to do the same. The parts of the

Bible that I always found the most arduous to memorize were the genealogies—Jehoram begat Uzziah, Josiah begat Jeconiah. As a child, I wondered what could possibly be the point of these long lists of names?

I now see that the authors of Biblical genealogies were motivated by the same impulse that motivates twenty-first-century customers to sign up for services like Ancestry or FamilyTree or MyHeritage, which combine genetic testing with archival records to construct people's family trees going back generations. It is the same impulse that motivates the Mormon church to record the names of the dead and store them in a Utah repository that could withstand a nuclear bomb.[2] Knowing who begat whom can inform and legitimize identity. Genealogy can foster solidarity and a sense of belonging: "We are family."

Who do you think of when you think of family? I think of my children and their father, my brother and mother and father and stepmother and half-brother and stepsisters, my brother's wife and the children they might have one day, my father's three siblings and my four first cousins on his side and their children, of my mother's three siblings and my four first cousins on her side and their children. My emotional ties and my genetic relationships with these people vary from strong to nonexistent. But they are all my family.

As you go back in time, your list of family members quickly becomes sprawling, as your number of ancestors doubles with every generation—two parents, four grandparents, eight great-grandparents, etc., etc. Going back 33 generations, or about 1,000 years, yields $2^{33} = 8,589,934,592$.

There weren't even 8 billion people *alive* back then, but some people are your ancestors multiple times over. My uncle Sean and my aunt Kristin, for example, are "kissing cousins" whose grandfathers were brothers. So my cousin Sterling doesn't have $2^4 = 16$ great-great-grandparents; he has, instead, 14, with one pair of those great-great-grandparents popping up twice in his family tree. Your pedigree, the canopy of your family tree, starts collapsing on itself.

The fact that humans sometimes reproduce with their cousins is only one of the many processes that makes human mating, well, complicated. People have historically stayed put, more or less—finding a mate and making babies not far from where their parents made babies. Migrations, particularly across long distances, were rare throughout human history. Even today, in the age of jet planes and interstate highways, the typical American lives only eighteen miles from his mother.[3] And among their geographically close neighbors, people mate with other people who share their language, their culture, their social class.

Because sex is complicated, it took a bit of complicated math to estimate how long ago in human history was the most recent common ancestor of *all* humans, i.e., someone who is in the family tree of everyone alive now. And the answer is—not that long ago: within the last few thousand years.[4] One conservative estimate is around 1500 B.C., as the Hittites were learning how to forge iron weapons. But it could be as recent as around 50 A.D., right around the time that Nero fiddled as Rome burned. Go back a little further, to sometime between 5000 and 2000 B.C., as the Sumerians were developing a written alphabet and Egypt's first dynasty was being established, you reach an even more remarkable point—*everyone* alive then, if they left any descendants at all, was a common ancestor of *everyone* alive now. If we trace our family trees back far enough, they all become one and the same.

This might seem impossible, as most of our recent ancestors lived and died in a single place, separated by massive un-navigable distances from most of the rest of the globe. To understand this finding, we need to appreciate just *how many* people we are talking about, when we are talking about the number of ancestors you had a thousand years ago. Even rare events, like long-distance migrations and matings between people divided by language and culture and class, were bound to happen every once in a while, in every family tree.

And those rare events tie you to the rest of the globe. Ultimately your family is my family, and my family is your family. As the population geneticist Graham Coop summarized, "Your family tree is vast and vastly messy; no one is descended from just one group of people." We all have a legitimate claim to our identity as part of a shared human family. "We're all descended from everyone in the world."[5]

Genealogical versus Genetic Ancestors

Let's complicate things a bit more: what I just described are your genealogical ancestors; however, your *genealogical* ancestors are not necessarily your *genetic* ancestors, particularly when we go back many generations. Excluding the sex chromosomes (X or Y), you inherited 22 chromosomes from your father. In making the sperm cell that became you, the chromosomes that your father got from *his* mother and father traded chunks of genetic material, in order to produce a new DNA sequence that is uniquely yours. There are, on average, 33 of these *recombination events* that occur every time a genome is transmitted to the next generation. So, the 22 chromosomes that you inherited can be broken down into $22 + 33 = 55$ different chunks, each of which can be traced back to one of your two paternal grandparents.

The same process happened, of course, in the previous generation, so you could also break down the chromosomes you inherited into $22 + 33 \times 2 = 88$ chunks, each of which can be traced back to one of your four paternal great-grandparents. This early on in your family history, the number of chunks is way larger (88) than the number of genealogical ancestors (4), so you are almost guaranteed to have inherited DNA from these people.

But the numbers start shifting quickly. Going back 42 generations, like from Jesus to Abraham, means that your DNA can be broken down into $2 \times (22 + 33 \times 41) = 2{,}750$ chunks, each of which can be traced to one of $2^{41} = > 2$ *trillion* ancestors. Obviously, you don't have 2 trillion ancestors; some of your ancestors are your ancestors many times over. But you have a lot—way more than you have chunks of DNA. So the chances that DNA from any one specific genealogical ancestor from nine generations ago still lurks in your genome is exceedingly small.

The fact that we did not inherit any DNA from the vast majority of our distant genealogical ancestors helps us make sense of two truths that might otherwise seem paradoxical. First, going back just a few thousand years, everyone's family tree converges, regardless of where they live in the world; we are all descended from everyone in the world. And second, people who live in different parts of the world are genetically different from one another, and these genetic differences can be very old, much older than a few thousand years.

One of your genealogical ancestors might have been part of an occupying military force, and his rape of another one of your genealogical ancestors links your family tree to the family tree of people who now live on the opposite side of the globe. Chances are, though, that none of that man's DNA still lurks in your genome (again, the Y chromosome is an exception here). This is because the chances that any one of our distant genealogical ancestors is also a genetic ancestor are small. Genealogical ancestors are regularly lost from your DNA—and with them go your genetic connections to distant peoples.

You only inherited DNA from a tiny fraction of your pool of genealogical ancestors. And most of your ancestors lived and died and reproduced in geographical proximity to one another. And most mating opportunities are limited not just by the need to be in close physical proximity, but also by complex cultural rules about who is sexually available to whom. The net result of this process is that genetic variation in humans has *structure*. That is, there are patterns in how the genetic make-up of any one person resembles, and diverges from, the genetic make-up of anyone else. These patterns—this structure—reflect both geography and culture.

Ancestry versus Race

The largest patterns of genetic similarities and dissimilarities among humans reflect the largest geographical barriers and boundaries—seas and oceans and deserts and continental divides. People whose genetic ancestors resided in East Asia are more genetically similar to one another than they are to people whose genetic ancestors resided in Europe. Statistical patterns of genetic similarity and dissimilarity can, therefore, be summarized by grouping people based on their genetic resemblance and then attaching labels to those groups that involve references to continental geography (Africa vs. Asia vs. Europe). This custom is sensible enough. The phrase

"African ancestry" works as a scientific shorthand to describe a group of people who are genetically similar to one another because they share many genetic ancestors and those genetic ancestors lived on the continent of Africa.

But as scientists made progress in learning to how analyze patterns of genetic similarity and dissimilarity across human populations, and began to attach geographical labels to clusters of people who shared genetic ancestry, others began to sound an alarm. Were geneticists reinventing race as a biological reality rather than as a social construction? This prospect was alarming because biological conceptions of race have long been used to justify oppression. In her book *Fatal Invention: How Science, Politics, and Big Business Re-Create Race in the 21st Century,* Dorothy Roberts pointed out that "making race a biological concept" has always "served an important ideological function": "Treating race as biology constituted the only suitable 'moral apology' . . . for slavery in a society that claimed equality as its most cherished ideal."[6] Arguing that the biological concept of race is "problematic at best and harmful at worst," Roberts and three colleagues went on to call for scientists to use terms like "ancestry" and "population" rather than "race."[7]

This distinction between ancestry and race is sometimes dismissed as a bit of sophistry that allow scientists to get away with talking about biological difference between races without actually using the *r*-word. For instance, in a podcast interview, Charles Murray, of *The Bell Curve* fame, tossed off the comment that "The word 'populations' is what the geneticists like to use now instead of race, and I don't blame them."[8] Comments like these collapse race and genetic ancestry into a single idea, and in so doing, perpetuate the idea that racial inequalities in outcomes must be due to innate biological differences. It is, therefore, critical to understand why this collapse is a mistake.

In the 1995 movie *Clueless*, the main character, Cher Horowitz, refers to one of her social rivals as a "full-on Monet"—"It's like a painting, see? From far away, it's okay, but up close it's a big ol' mess." The analysis of the genetic structure of human populations is a full-on Monet. Zoom out, and the pattern looks clear enough, with superpopulations corresponding to major continents.

Zoom in closer, and you can still see patterns, although they start to look muddier. A study of how "genes mirror geography" in Europe did, indeed, find that people whose grandparents all resided in what is now France are more genetically similar to one another than to people whose grandparents all resided in what is now Sweden. But some Italians cluster, genetically, away from other Italians, who overlap with people whose grandparents are all Swiss. And Jews are nowhere to be found in the sample. Neither is anyone whose grandparents didn't all hail from the same location.

Zoom closer in on an individual person, and the clarity you had from a distance dissipates. Wherever there are boundaries, there are people whose histories stretch across those boundaries. Particularly when people's family histories have been shaped by colonialism or enslavement or occupation or migration or war, by the often forcible bringing together of previously separated peoples, they defy easy categorization into the patterns that seemed so readily apparent from a distance.

Nevertheless, just as we might refer to one part of a Monet painting as "the sky," we refer to one part of the human population as "European ancestry." Or, more narrowly, "Northern European ancestry." Or more narrowly still, "White British ancestry."

If patterns of genetic ancestry among humans are a "full-on Monet," then our racial distinctions are more like a Mondrian painting— brightly-contrasting primary colors separated by clear boundaries. Racial categories are discrete and mutually exclusive: the US Census, which began in 1790, did not allow people to pick more than one race until 2000. And these categories are inherently hierarchical, as the process of racial categorization serves to restrict who has access to power, wealth, and physical space. Audrey and Brian Smedley wrote in their summary of anthropological and historical perspectives on race, "Race essentializes and stereotypes people, their social statuses, their social behaviors, and their social ranking."[9]

To be clear, I'm not claiming that race and ancestry are totally independent. *Of course* racial groups differ in genetic ancestry, and this correspondence is shaped by the social and legal history of racial classification. In the United States, for instance, laws that mandated racial segregation (e.g., "Whites only" schools, drinking fountains, train cars, swimming pools, and other spaces) required an explicit definition of who was White and who was not. In the early twentieth century, several states in the American South passed racial categorization laws with a "one-drop" rule: any amount of non-White parentage was sufficient to negate someone's status atop the American racial hierarchy. Virginia's Racial Integrity Act of 1924 put it this way: a person was White if she has "no trace whatsoever of any blood other than Caucasian." One-drop rules are an extreme version of "hypodescent" rules, where the children of mixed-race unions are assigned the racial category of the parent with the socially subordinate race.

Because of the social and legal history of hypodescent rules for racial classification, people who currently identify as White in the United States are very unlikely to have *any* amount of non-European genetic ancestry. One study estimated that only 0.3 percent of people who self-identified as White had any African ancestry. At the same time, people from Africa were forcibly brought to the US and enslaved, and their descendants are categorized as "Black"— and so nearly all people who are socially categorized as Black (99.7 percent, according to one study) have at least some "African" genetic ancestry.[10]

Yet despite this nearly 1:1 correspondence between having exclusively European genetic ancestry and being racially categorized as White, or between having some African genetic ancestry and being racially categorized as Black, it would *still* be a mistake to conceptualize race as being synonymous with ancestry—for four reasons.

First, when we categorize people into races, we elevate some distinctions between people while eliding others. These racial distinctions are culturally and historically contingent. If you look back at the work of early-twentieth-century eugenics thinkers, the "racial" questions that they were obsessed with might strike you as bizarre. Read Carl Brigham, a Princeton psychology professor and an early proponent of intelligence testing, on "the race question," and he is trying to figure out how much Nordic, Alpine, and Mediterranean "blood" immigrants from each European country have.[11] Wave after wave of European immigrants—Italians, the Irish, Jews—were not initially considered part of the "White" dominant class in the United States.[12] This social contingency of how race is defined is inescapable, because race (unlike ancestry) is an inherently hierarchical concept that serves to structure *who* has access to spaces and social power.

Second, whether people are socially categorized as different races does not correspond to their degree of ancestral genetic difference in any straightforward way. African ancestry populations, in particular, are remarkable for their genetic diversity, with some African groups being more different from each other than Europeans are from East Asians. Yet everyone of African descent in America is folded into the same category of "Black." Similarly, with regard to genetic ancestry, people of South Asian descent are distinguishable from people of East Asian descent, so much so that South Asian is commonly considered its own continental super-population.[13] Yet the US Census Bureau defines the racial category of Asian as "a person having origins in any of the original peoples of the Far East, Southeast Asia, or the Indian subcontinent."[14]

Third, there can be a range of continental ancestral backgrounds within any one self-identified racial group, making it nearly impossible to ascertain the ancestry of a single person with any degree of confidence if all you have is information about race. As we've seen, nearly 100 percent of self-identified Black Americans have some African ancestry (which itself is a very heterogeneous supercategory), but at the same time over 90 percent of Black people in America also have some European ancestry.[15] And while you can be fairly confident that someone in the United States who identifies as White will have European genetic ancestry, the reverse is not true— people who have *some* amount of European ancestry could identify as nearly any other racial category.

Fourth, ancestry can be quantified very granularly, and these granular distinctions are impossible to describe using the familiar language of race. As I mentioned earlier, "one-drop" social rules have guaranteed that Americans who identify as being White are very unlikely to have any genetic ancestry that that is not European, so in this case self-reported race and genetic ancestry appear to converge. But even within a population of exclusively European-ancestry individuals, there is *still* genetic structure that reflects finer geographical gradations, along with all the potential differences in language and culture and class that make mating non-random.

To capture this structure—which is potentially "cryptic," or hidden—geneticists commonly use a method called "principal components analysis" (PCA). PCA analyzes patterns of genetic similarity between people (which reflects them having ancestors in common) and produces a set of ancestry-informative principal components, or "PCs."[16] Each of these variables is continuous (meaning that people can be high, low, or anywhere in between), rather than a yes-or-no category. And it's not uncommon for researchers to include 40 or more ancestry-informative PCs in their study, even when they are focusing only on a group of people that, on paper, already looks quite homogenous (e.g., people who all identify as "White British").[17] This approach finely characterizes a group of people who would otherwise be all lumped together into a single racial category, and each ancestry-informative PC would be impossible to interpret in racial terms.

Putting all of this together, Roberts and her colleagues summarized the distinction between race and ancestry like this: "Ancestry is a *process*-based concept, a statement about an individual's relationship to other individuals in their genealogical history; thus, it is a very personal understanding of one's genomic heritage. Race, on the other hand, is a *pattern*-based concept that has led scientists and laypersons alike to draw conclusions about hierarchical organization of humans, which connect an individual to a larger preconceived geographically circumscribed or socially constructed group" (emphasis mine).[18]

Why Ancestry Matters for GWAS

Historically, scientific racists would point out differences between the skulls of people who had been assigned to different races. Modern-day scientific racists are more likely to talk about patterns of genetic ancestry to make the case for inborn racial differences. But, as I've described so far in this chapter, a closer look at the science of genetic ancestry makes it clear that "race does not stand up scientifically, period."[19] Genetic data has not "proved" the biological reality of race. Instead, in an ironic twist, understanding how socially defined racial groups differ in their genetic ancestry helps us see why modern "race science" is actually pseudoscience. In the next section, I'll describe how genetic differences between different populations complicate the effort, often ill-intentioned, to take what GWAS has discovered about individual differences *within* a population and draw conclusions about the source of *between*-population differences.

The first way that populations differ is in which genetic variants are present and how common they are. A genetic variant that is rare in one population could be common in another.[20] About three-quarters of genetic variants are found only in a single continental group, or even in a single sub-continental group, with African ancestry populations showing the greatest genetic diversity. As a consequence, the genetic variants that are most important for a phenotype in one population are not necessarily the most important in another population: a particular mutation in the *CFTR* gene, for instance, is responsible for over 70 percent of cystic fibrosis cases in European ancestry populations but less than 30 percent of cases in African ancestry populations.[21] Studies that represent more of the world's genetic diversity, therefore, hold extraordinary potential to discover new genetic variants that would never have been discovered in work focused only on European ancestry populations. A study of Africans living in Ethopia, Tanzania, and Botswana, for example, discovered new genetic variants affecting skin pigmentation, which varies widely across the African continent, from the light-skinned San to the very dark-skinned Nilo-Saharans in East Africa.[22]

The second way that populations differ is in the pattern of linkage disequilibrium (LD) seen in the genome, i.e., in which genetic variants are correlated with which other variants. Again, African ancestry populations are particularly noteworthy here, as LD is lower than in non-African ancestry populations and is heterogeneous across different African populations.[23] Recall that a GWAS typically does not measure every single DNA letter, but rather a small subset of SNPs. The results of a GWAS, then, can be driven by associations with the measured SNP itself *or* by associations with any genetic variant in LD with the measured SNP. The devil is in this technical detail: the exact same genetic variant could be associated with an outcome across populations, but the results of a GWAS conducted in just one population *still* might not translate, because the SNPs that are actually measured "tag" that causal variant differently in one population than in another.

The bottom line, then, is that we cannot and should not expect GWAS results to be "portable" across genetic ancestries *or* socially defined races. What you discover in one group isn't expected to apply to another group, and if you study a different group, you might discover different genes. This expectation is clearly borne out by the data. Looking across a diverse set of phenotypes ranging from HDL cholesterol to schizophrenia, polygenic indices based on analyses of European ancestry populations are less strongly related to phenotypes measured in other populations, particularly African ancestry groups.[24] When researchers have used an educational attainment GWAS to construct a polygenic index in White-identified, European-ancestry samples from the UK or Wisconsin or New Zealand, then the score "worked"—it captured more than 10 percent of the variance in educational attainment in those samples. But when researchers tested the polygenic index in a sample of African Americans, who are all expected to have at least some African ancestry, it was much less strongly associated with educational attainment.[25] As we've seen with studies of genetically simpler phenotypes like cystic fibrosis or skin color, future genetic studies of educational attainment or other social and behavioral phenotypes in people from Africa and the African diaspora might discover different genes than are relevant in European populations.

Eurocentric Bias of GWAS Research

For now, however, nearly all GWAS research has focused on people whose genetic ancestry is exclusively European. As of 2019, people of European descent made up only 16 percent of the global population but accounted for nearly 80 percent of GWAS participants. This situation is not improving, despite the falling cost of genotyping. In the last five years, the share of genetics research focused on people of European ancestry has held steady, even as the overall number of genotyped people continues to explode.[26]

Because results from genetic research conducted in one ancestry group are not expected to be portable or generalizable to people from another ancestry group, the Eurocentrism of current genetic research has the potential to exacerbate existing health disparities.[27] Work in medical genetics is developing polygenic risk scores to predict the future onset of cancer, obesity, heart attacks, and diabetes. The goal of such work is to catch people at high risk earlier and match them to effective treatments more quickly. These same chronic diseases, however, disproportionately affect people of color in the United States. The use of polygenic indices, therefore, has the potential to widen health disparities even more, by improving health outcomes only for those people who are exclusively of European descent.

The only way to surmount this problem is to invest preferentially in genetic research on the rest of the global population. But genetics research does not just disproportionately study White people. It also is disproportionately conducted by White people. The collection and analysis of genetic data from populations of non-European ancestry thus presents a double bind. Without conducting genetic research with the entire global population, there is a danger that genetic knowledge will only benefit people who are already advantaged. But there are valid and deep-seated concerns that DNA will become yet another valuable resource extracted from marginalized populations, by White people for the benefit of White people, while leaving participants vulnerable to greater surveillance, discrimination, and other harms. In this way, the Eurocentrism of genetic research is an example of how racist systems can interact and reinforce each other, making it difficult to change any one system in isolation.

Ecological Fallacies and Racist Priors

By this point, it is hopefully becoming clear why any claims about "genetic" racial differences in intelligence or educational attainment or criminality or any behavioral trait are scientifically baseless.

Because existing large-scale GWASs are based on European ancestry populations, our knowledge about how genetics are related to inequalities in life outcomes is *entirely* about individual differences between people whose ancestry is entirely European and whose self-identified race is likely White. We can't assume that genetic associations are working the same way in people with different genetic ancestry, in part for fairly technical reasons concerning how the genome is measured and structured. We can't "compare" the genetics of different ancestry groups using their polygenic indices. We can't assume that everyone who has the same race shares the same genetic ancestry. Whether we are talking about complicated social phenotypes like education or relatively uncontroversial physical ones like height— modern molecular genetic studies, like the older twin studies, have told us a whole lot of nothing about the causes of racial inequalities.

But even in the complete absence of any genetic "evidence" for genetic racial differences, people commonly mount an argument that, on its face, might sound reasonable. If (1) individual differences in education *within* White populations are caused by genetic differences, and (2) Black people in the US, on average, have lower levels of education, then (3) doesn't it just make intuitive sense that the group difference is also caused—at least a little bit—by genes?

It might seem simple to make the leap from "X causes individual differences *within* a group" to "differences in average levels of X cause average differences *between* groups." But from a statistical perspective, assuming that correlations within a group tell you something about the causes of between-group differences is a leap that only fools would make. It is an ecological fallacy.

It's helpful to explain the ecological fallacy in the context of an example that has nothing to do with genetics. This will make it clearer, hopefully, that my objection to connecting individual differences to group differences is not simply motivated by the fact that I find the conclusion that racial group differences are "genetic" to be unpalatable. I am not brandishing a politically motivated talking point. I'm making a statistical point that applies *anytime* we are trying to jump from one level of aggregation to another—not just when we are talking about the inflammatory topic of genes and race.

A pioneering paper on the ecological fallacy was written in 1950 by a sociologist, W. S. Robinson, who presented two sets of correlations.[28] The first was the individual correlation between being foreign-born versus native-born (to the US) and being illiterate in English. It was positive (~.12): adults who were born outside of the US had more difficulty reading and writing fluently in English than people born here. Next, he calculated what he termed the "ecological" correlation between the percent of foreign-born residents in a state and the corresponding illiteracy rate in that state. It was not only different from the individual correlation; it actually *switched direction* (~ -.5). The explanation for this apparent paradox is that immigrants to the United States tend to settle in states where the native-born residents are more likely to be literate. That is, states differ in their literacy outcomes for lots of reasons *other* than the variable that we measured when calculating the individual correlation.

Now imagine that we only had *some* of the information that Robinson had. In this scenario, we can observe the positive individual correlation between foreign-born status and being illiterate in English, but we can only observe this individual correlation in *some* of the US states. It's a decent guess that the individual correlation is the same in the other states, but it's just that—a decent guess. And, in this scenario, we can observe that American states differ in their rates of illiteracy. But, because of measurement problems, we don't have any data on how the states compare in their percentage of foreign-born residents. We can see the group differences in the outcomes, but we can't measure group differences in the putative explanatory variable.

Given this partial information, one could—incorrectly—assume that the ecological correlation was going to be identical, or at least in the same direction, as the individual correlation. So you'd look at the states that had higher illiteracy rates and conclude that they had more foreign-born residents. You might hedge a bit: "Well, I'm not saying the *only* reason that Mississippi is more illiterate than California is that Mississippi has more immigrants. Maybe that's only half the reason."

Except, of course, you would have it exactly backwards—the states that you observed as having higher illiteracy rates actually have fewer foreign-born residents.

This situation, characterized by incomplete information and faulty assumptions, is the exact situation we find ourselves in in the context of between-population differences in almost any trait, whether it be height or intelligence test scores or educational attainment. We can observe the positive correlation between genetic variants and educational attainment, for instance, within one group (European ancestry). We can observe differences in educational outcomes between ancestry groups. We can't reliably measure group differences in the putative explanatory variable—genes. Within one group, we can observe an individual-level correlation between genes and educational attainment. It might seem only rational to infer that this means that one ancestry group has worse educational outcomes because whatever genetic variants cause better outcomes in education are rarer.

But in reality, you could have it exactly backwards, and the genes that matter for education could be *more* common in the ancestry group with *worse* educational outcomes. The individual correlation and the ecological correlation are not just different things. *One of them does not give information about the other.* More than half a century later, Robinson's tersely written conclusion to his paper on ecological correlations remains relevant: "The only reasonable assumption is that an ecological correlation is almost certainly not equal to its corresponding individual correlation."

What, then, can we make of claims that "science says" racial disparities in life outcomes are due to genetic differences between the races? Yes, socially constructed race differences are systematically related to genetic ancestry. And, yes, within European-ancestry populations, genetic differences between people are associated with differences in their socially important life outcomes. But neither one of these pieces of information gives you any information about the sources of racial disparities.

In Bayesian statistics, there is a something called a *prior*, which is a mathematical representation of what you believe—and how uncertain you are about those beliefs—before (prior to) any evidence is taken into account. What do you know—or believe that you know—when no information is available? That is the situation we find ourselves in regarding between-population genetic differences in complex life outcomes such as education. Take that one step further: What are those prior beliefs based on, if it's not scientific evidence?

The prior belief that White people enjoy better life outcomes because of their genetics is a perniciously persistent one. In the 1960s, the educational psychologist Arthur Jensen speculated that the educational progress of Black schoolchildren would not be improvable beyond a certain point, and certainly not to the level of White schoolchildren, because of the limits imposed by genetics.[29] In the 1990s, Herrnstein and Murray blithely presented their hypothesis that at least part of the reason that Black and Hispanic people in America had lower average IQ test scores than White people was because of the genetic differences between them.[30] Today, the "race realist" and "human biodiversity" communities post copies of *Nature Genetics* articles that they believe make the case that there are genetic differences between races that cause differences in intelligence test scores, impulsive behavior, and economic success.

These communities insist that they are "just asking" an empirical question—are there genetic differences between genetic ancestry groups that cause differences in average life outcomes? But once we appreciate that their hypothesis is not grounded in any legitimate scientific evidence, we can recognize that the question is founded on a racist prior belief about the supremacy of certain racial groups.

Antiracism and Responsibility in a Postgenomic World

I've described why speculation about genetically based racial differences is not grounded in science today—but what about tomorrow? After all, the field of human genetics is progressing at an extraordinary pace. Already, researchers in the field of population genetics are using GWAS results to try to understand how differences between humans might have evolved over time.[31] In 2018, the geneticist David

Reich wrote an opinion piece in the *New York Times* urging people to consider: "How should we prepare for the likelihood that in the coming years, genetic studies will show that many traits are influenced by genetic variations, and that these traits will differ on average across human populations? It will be impossible—indeed, anti-scientific, foolish and absurd—to deny those differences."[32]

The writer Sam Harris made a similar argument when I appeared on his podcast in the summer of 2020. After explaining why I thought that the idea of "genetic" differences in intelligence between racial groups was not an idea supported by science (for reasons I've described in this chapter), Harris pushed back, arguing that my position "would be bowled over by coming developments in genetics and other sciences." He predicted that "if you could list the top 100 things we care about in human beings, intelligence would be one . . . it would be a miracle if the mean value for the 100 things we care about would be the same for every conceivable group of human beings. . . . So, my view, politically, is we need to be able to absorb that fact."[33]

I am skeptical of Reich's and Harris's premise, for multiple reasons. I am skeptical that differences between groups in social and behavioral traits like education will ever be best understood at the genetic level of analysis. (I return to the idea of levels of analysis in chapter 8.) I am skeptical that scientific results will just so happen to turn out to be consistent with the "Just So" stories told by White people crafting a "moral apology" for slavery and oppression. I am skeptical more generally of anyone who claims clairvoyance about what the science of human genetics will tell us, as it has heretofore been full of surprises.

But while I am skeptical of their priors about what future analyses of genomic data will find, I do agree with them on one point: people's moral commitments to racial equality are on shaky ground if they depend on exact genetic sameness across human populations. Consider, for example, Ibram X. Kendi's best-selling book, *How to Be an Antiracist*.[34] In his chapter "Biology," Kendi insisted that a "biological antiracist" was "one who is expressing the idea that the races are meaningfully the same in their biology and *there are no genetic racial differences*" (emphasis added).

As I discussed above, race is not a valid biological category. But to hold that there are no genetic differences between groups of people who identify as different races is simply incorrect: as I described previously in this chapter, racial groups differ in genetic ancestry, and so differ in which genetic variants are present and how common those variants are. Must our commitment to antiracism and racial equality be built on such tottering foundations? As I told Harris, "it is a grave mistake to stake claims for equity, or inclusion, or justice . . . on the absence of genetic differences." To do so is to make our moral commitments fundamentally unstable, potentially toppled by the next paper in *Nature Genetics*.

If we are to make our commitment to antiracism stable in a post-genomic world, I think it is necessary, however unpalatable, to consider Reich's question about how we should prepare for scientific discoveries, whatever they might be. Let us not flinch from considering what seems like the worst-case scenario: What if, next year, there suddenly emerged scientific evidence showing that European-ancestry populations evolved in ways that made them genetically more prone, on average, to develop cognitive abilities of the sort that earn high test scores in school? How would we "absorb" that fact?

Reich answered his question with a plea to "treat each human being as an individual" and "accord [each person] the same freedoms and opportunities regardless of those differences." While avoiding stereotyping on the basis of group identity and affording equal opportunity are goals I agree with, I don't think these steps go far enough to address his question. Too often, the idea of "equal opportunity" acts as a rhetorical dodge, a way to avoid reckoning with profound inequalities of outcome. Treating everyone exactly the same right now cannot help but reproduce inequalities of the past.

Instead, if we are interested in making our commitment to racial equality "genetics-proof," I think we must dismantle the false distinction between "inequalities that society is responsible for addressing" and "inequalities that are caused by differences in biology."[35]

The mistaken idea that genetic causes operate as a boundary for social responsibility was evident in Sam Harris's comments toward the end of our podcast conversation. We were speaking in the middle of a summer rocked by the murders of George Floyd and Breonna Taylor at the hands of police, with Black Lives Matter protests happening in cities around the world. The books *White Fragility* and *So You Want to Talk about Race* topped the *New York Times* bestseller list,[36] signs of a national conversation that was happening about racial disparities in policing, housing, health care, education, wealth, and political power in America. About which Harris asked:

> The real question is what is the *cause* of all of these disparities? The problem politically at the moment is, when you're talking about White-Black differences in American society . . . the only acceptable answer in many quarters to account for these differences is White racism, or systemic racism, institutional racism, some holdover effect from slavery and Jim Crow. . . . It is deeply unstable, because we will find out things, about differences among groups.

These comments invoke an either/or: it is either systemic racism, which White people presumably have a moral responsibility to address, *or* genetics, which is presumed to be a fixed and deterministic aspect of biology for which no one should be made to feel responsible. As the feminist philosopher Kate Manne put it in her work on sexism, "The unstated premise here is a version of the 'ought implies can' principle—possibly weakened to something like 'can't even implies don't bother.'"[37] The unstated premise of positioning genetics, as *opposed* to racism, as a cause of racial disparities is to imply that people, particularly White people atop a racial hierarchy, need not feel morally compelled to *do* anything about changing disparities if their cause is genes.

The crucial flaw at the heart of this thinking is *not* that it posits the existence of genetic differences between racial groups. As we've seen, race is a poor representation of genetic ancestry, but it is not unrelated to genetic ancestry. Nor is the crucial flaw that it links genetic difference with differences in life outcomes. As I'll explain in this book, there is a plethora of scientific evidence that our genes matter for shaping our selves, not just for our physical characteristics.

The crucial flaw in this thinking is that it presumes that the existence of genetically caused human differences waives our social responsibility to address inequality. As I will describe in the upcoming chapters, the existence of genetic influence—regardless of how it is distributed across socially defined groups—does not impose a hard boundary on the prospect of social change via social mechanisms, nor does it operate as a "get out of jail free" card for our social responsibilities.

Ultimately, I think it is likely that the upcoming avalanche of genomic data from multi-ancestry populations will show that populations differ minimally, if at all, in the prevalence of genetic variants relevant for psychological characteristics, like the cognitive abilities tested by our current battery of intelligence tests. But, no matter how people differ genetically, no matter how those genetic differences between people are distributed across socially defined racial groups, no matter how strongly those genetic differences influence the development of human characteristics, no matter whether those characteristics are physiological or psychological, we are *still* not absolved of the responsibility to arrange society to the benefit of all people, not just the tiny slice of global genetic diversity that is people of predominantly European ancestry. And that responsibility must be lived out in our policies. That is, our policies should reflect the truth that, as the evolutionary biologist Theodosius Dobzhansky wrote, "genetic diversity is mankind's most precious resource, not a regrettable deviation from an ideal state of monotonous sameness. . . . Nonfulfillment of human potentialities is a waste of human resources."[38]

Summing Up, Looking Forward

Here, let me recapitulate the points I have made in this chapter: Genetic ancestry is a process-based concept that links people to their personal genealogical histories, whereas race is a pattern-based concept that links people to socially constructed groups in order to maintain hierarchical power relationships. GWAS research is examining individual differences in intelligence, behavior, and attainments *within* samples of people who all have exclusively "European" ancestry and who, because of the way Whiteness is socially constructed in the United States, would all likely identify as White. These results don't necessarily generalize to other ancestral populations and cannot be used to make comparisons between ancestral populations. There is no scientific evidence for genetically based differences in intelligence test score performance between racial groups or between ancestral populations. Rather than based on scientific evidence, the idea that Black people in America experience worse social outcomes because of their genetics is based on centuries of racist thought, which views differences between people in terms of a racialized hierarchy that has White people at the top of it. And, most crucially, our responsibility to arrange society so that it benefits all people, not just people with a certain set of genetic characteristics, is not obviated by any genetic discoveries.

It can be very difficult to keep these ideas clearly in mind when we discuss the relationship between genetics and social inequality. This difficulty is not an accident. This difficulty is the result of decades of racist thought that has persistently appropriated biology as part of its ideological toolkit for legitimizing a racial hierarchy. In the chapters to come, then, as I make the case for why we should take genes seriously as a cause of individual differences, it might be helpful to refer back to this chapter, to remind yourself why individual differences and racial differences *are not the same thing*, and why the existence of genetic differences does not obviate our social responsibility.

At the same time, this chapter has begun to touch on some important issues that need to be unpacked more slowly, like the idea that genetic causes can have social mechanisms (chapter 7), and the idea that genetic influence does not impose a hard boundary on the possibility of social change (chapters 8 and 9). But before we get there, we must grapple with a more basic issue: What is a genetic cause? As I described in the previous chapter, a GWAS correlates small bits of DNA with an outcome, but, as is the common refrain—correlation does not equal causation. How do we get from the correlational results of GWAS to an understanding of how genes may be a *cause* of social inequalities in a particular historical and cultural context? Answering that question, in turn, requires us to be precise about our definition of the word "cause," and it is to that topic that we turn our attention in the next chapter.

5

A Lottery of Life Chances

Every Psychology 101 student knows that "correlation does not equal causation." Restaurants that add more grated sea urchin to every dish might be rated higher on Yelp, but that correlation does *not* mean that adding sea urchin to every menu is going to cause people to enjoy restaurants more. Similarly, genome-wide association studies have found that European-ancestry children with certain genetic variants do better in school, but does that mean that those genetic variants have *caused* people's educational outcomes?

This question is more complicated than it might appear at first glance, because in order to answer it, we have to wrestle with a bigger question: What does it mean to be a cause? As I will describe in this chapter, the word "cause" has never had a single definition. And people's definitions of the word "cause" are especially mercurial when the question at hand is whether genes can be causes. Poke in one direction or another, and people's definitions expand or contract, as need be, to encompass the things they want to embrace as causal, and to evade others. In order to make the argument that genes *cause* social inequalities in income and education and health and well-being, we need to be specific about what a cause is—and what it isn't.

An Adoption Experiment

In 1966, the Communist government of Romania outlawed abortion and contraception for women who were under age forty-five or who had fewer than five children.[1] Forced to bear children they didn't want and couldn't feed, women gave up their babies to state-run orphanages in droves. More than 500,000 children were raised in state-run institutions—a "lost generation" raised in "slaughterhouses of souls."[2] When the authoritarian government fell and Romania was opened to the West, visitors to the country's orphanages were horrified at what they found. Hundreds of children were held in barren metal cribs, sitting in eerie quiet. In the absence of any consistent attachment to a caregiver, subjected to daily violence and humiliation, and despairing of their emotional and intellectual needs ever being met by another human, the children had retreated into silence.

Upon seeing the extreme neglect that Romanian orphans were experiencing in state-run institutions, a group of US scientists saw an extraordinary opportunity to answer a question about human psychology. Is there a critical window during which a sufficiently good environment is necessary for normal psychological development? In the mid-twentieth century, the psychologist Harry Harlow cruelly and unethically experimented with young monkeys, whom he had separated from their mothers, in order to address this question. Harlow's barbaric experiments provided a haunting demonstration of how young primates need not just food and milk, but also physical closeness from a caregiver, in order to thrive.[3] In the following decades, John Bowlby, a British psychoanalyst, and then his student, the psychologist Mary Ainsworth, extended Harlow's insights into a theory about what they called "attachment."[4] More than just physical proximity, young primates need a warm and responsive relationship with a caregiver to mature cognitively and emotionally.

Now, here was a group of children who had been deprived of that attachment relationship. Was the theory right? Could children recover if they were rescued from the deprived environment? Did it matter how early you intervened?

In order to address this question, the scientists created a foster care system in Romania where none existed. Then they literally drew children's names out of a hat to determine which ones would stay in the orphanage and which ones would live with foster families. It was a lottery of life chances, set up to ask scientific questions.[5] Nothing separated the children who were raised in foster care from the ones who remained in the orphanage but sheer randomness. (The researchers obtained appropriate approvals to conduct this experiment, and they discuss the ethical issues raised by their study in at least one paper. The ethical acceptability of experimentation with vulnerable populations, however, remains controversial.[6])

Researchers have continued to follow both groups of children—the ones whose names were pulled from a hat, and the ones whose names weren't—and test them in a variety of ways to see how their bodies, brains, emotions, minds, and lives have diverged. In 2007, a landmark paper from the study was published in *Science*.[7] At age 54 months, the average IQ of children who had been randomized to foster care was 81. For children who had been randomized to stay in the orphanage, the average IQ was 73. (IQ scores are designed so that the average score in the population is 100 points and the standard deviation is 15 points. Based on these norms, an IQ of 81 is lower than about 90% of people, and an IQ of 73 is lower than 96% of people.) This difference was "significant," meaning a difference that was unlikely to occur by chance.

The conclusions of the study were straightforward: being raised in a family, rather than in a barren crib where no one holds you or talks to you or reads to you or lets you go outside, makes you smarter. Also, timing matters. Children who had been rescued from the orphanage at the youngest ages had the highest IQs, on average. In contrast, children who were not randomized to foster care until after age 30 months had average IQs no different from children who remained in the orphanage throughout their childhood. (Thirty months is still *really* young. Many children that age are still in diapers.)

You might think that I'm telling you about this study to prove that the quality of the environment, particularly the *early* environment, influences the development of cognitive abilities in early childhood. And, yes, the early environment certainly does affect cognition— but that's not my point. I want to direct your attention to a different question: Did being rescued from an orphanage and placed with a foster care family *cause* an increase in IQ?

The researchers themselves certainly thought so. They wrote in their paper: "We are confident that the differences [in IQ] that resulted from the foster care intervention reflect *true intervention effects*." (p. 1940, emphasis added). They were not claiming that foster care was *associated* with higher IQ or *correlated* with higher IQ. They were claiming that foster care, as compared to institutional care, *caused* an increase in IQ.

Among social scientists, their claim to have tested a causal effect is not particularly controversial. Most of us would interpret the results of a properly conducted experiment like this as evidence for a causal relationship. And our comfort with the word "cause" here implies a specific definition of what causation *is*: a cause is something that has made a difference.

Causes and Counterfactuals

In 1748, the Scottish philosopher David Hume[8] offered a definition of "cause" that was actually two definitions in one:

> We may define a cause to be *an object, followed by another, and where all the objects, similar to the first, are followed by objects similar to the second.* Or, in other words, *where, if the first object had not been, the second never had existed.*

The first half of Hume's definition is about *regularity*—if you see one thing, do you always see a certain other thing? If I flick the light switch, the lights regularly, and almost without exception, come on. From here on out, I'm going to refer to the thing we think is a cause (flicking the light switch) as **X**, and its effect (lights coming on) as **Y**.

Regularity accounts of causality occupied philosophers' attention for the next two centuries, while the second half of Hume's definition—*where, if the first object had not been, the second had never existed*—was relatively neglected. Only in the 1970s did the philosopher David Lewis[9] formulate a definition of cause that more closely resembled the second half of Hume's definition. Lewis described a cause as "*something that makes a difference*, and the difference it makes must be a difference from *what would have happened* without it" (emphasis added).

Lewis's definition of a cause is all about the *counterfactual*—**X** happened, but what if **X** had *not* happened? What if the child who was adopted into foster care *had not* been put into foster care? Under the counterfactual definition of the word "cause," to say that **X** causes **Y** is to say that, if **X** had not happened, then the probability of **Y** happening would be different. To say that foster care causes higher IQ is to say that if a child had not been adopted into foster care, then there is a chance that her IQ would be lower.

Lewis's paper might have been hailed as novel within philosophy, but the idea that causes are *difference-makers* has evolved, more or less independently, on multiple occasions. Here, for example, is John Stuart Mill (1843):

> If a person eats of a particular dish, and dies in consequence, that is, *would not have died if he had not eaten of it*, people would be apt to say that eating of the dish was the cause of his death (emphasis added).[10]

And, just one year after Lewis published his 1973 paper, the statistician Donald Rubin[11] defined causation in strikingly similar terms:

> Intuitively, the causal effect of one treatment, **E**, over another, **C**, for a particular unit and an interval of time from t_1 to t_2 is *the difference* between what would have happened at time t_2 if the unit had been exposed to **E** initiated at t_1 and what would have happened at t_2 if the unit had been exposed to **C** initiated at t_1: "If an hour ago I had taken two aspirins instead of just a glass of water, my headache would now be gone" (emphasis added).

Observing What Could Have Been

The 1998 movie *Sliding Doors* begins with this very question: "Have you ever wondered what might have been?" In an early scene, the main character, played by a pre-GOOP Gwyneth Paltrow, narrowly catches her train, arriving home to find her boyfriend in bed with another woman. In the next scene, Gwyneth Paltrow narrowly misses that same train, avoids discovering her boyfriend's infidelity, and continues to bumble along in a deeply disappointing relationship with a man who is, in her words, a "sad, sad wanker." The movie bounces back and forth between two alternate lives, between two potential outcomes—what would have happened if Gwyneth had or had not caught the train?

Counterfactuals are, generally speaking, just that—they are conditional statements about a world *that does not exist in fact*. This is what has been called the "fundamental problem of causal inference":[12] we almost never get to observe what might have been for a single individual. I do not get to see the *Sliding Doors*-esque alternative realities of my idiosyncratic life: What if I had taken that other job offer? What if I had accepted that other marriage proposal?

We also don't get to see the counterfactuals for the lives that we want to understand as scientists. A researcher cannot keep a child in an orphanage and also put that same child into foster care, and then compare the alternative realities of one child's life. You only have one life to live. You can't bake the same cake twice. You can't experience **X** *and* **Not-X**.

Often, the solution for the fundamental problem of causal inference is to compare outcomes *between* people who have experienced **X** and other people who have experienced **Not-X**. *Your* life after being rescued from orphanage hell tells me something about what *my* life might have been if I had also been rescued.

The obvious difficulty here is that your life will be different from my life even if we had both been put into foster care. You are you, and I am I—so how to isolate any differences in our lives to one element that you experienced and I didn't?

Experiments such as the Romanian orphanage study resolve that difficulty by studying *groups* of people and comparing their *average* outcomes, rather than the outcome for any one individual. In that particular example, 68 children were sent to foster care, and 68 remained in the orphanage. The idea is that averaging across 68 people who all have one thing in common—foster care vs. institutional care—averages out all the "noise" of their idiosyncratic life outcomes. All that remains is the "signal" driven by the thing they have in common.

But this works *only if* the thing that people have in common is the thing that researchers are interested in studying. If researchers had selected, for example, all the boys to remain in the orphanage and all the girls to go to foster care, there would be no way of telling if the statistical signal that is being detected is driven by being in foster care or being female. Part of the reason why every first-year undergraduate is told, at some point, that "correlation does not equal causation" is a variation on that point. Yes, volume of ice cream sales in a county are positively correlated with murder rates, but eating lots of ice cream isn't the only thing that those counties have in common—they also share being in warmer climates. Comparing groups of people in order to peer into a counterfactual world only works if you can isolate X from everything else that differs between people.

The need to isolate the putatively causal variable from everything else is why *randomness* is so important to experimental design. Usually, our life experiences are braided together. Random assignment, when researchers intervene in the universe and determine *who* experiences *what*, completely independent of all of their other life characteristics, untwists the braid. Children who got sent to foster care didn't get sent there because they deserved it, or because they were the tallest or prettiest or best-behaved or most in need of a loving home. They got sent there because their names were drawn out of a hat. Luck crashed into their lives, and that luck—*by virtue of being luck*—was isolated from the web of all the other causes of their life outcomes.

The statistical analyses conducted by the Romanian orphanage study were actually pretty simple. What was the average IQ of children who were adopted into foster care versus those who remained in the orphanage, and was the difference between those averages bigger than would be expected by sheer chance? How big the difference between the groups would have to be, and how worried we should be about falsely concluding that foster care "works" to raise IQ when it really doesn't, or falsely concluding that it doesn't work when it really does, are all important considerations for science. But these questions are beside the point here. The important point is this: an experiment where participants are *randomized* to different values of **X** (in this case, foster care versus institutionalized care) is generally accepted as a method that allows scientists to test whether **X** *causes* **Y** (in this case, increases in IQ) because the comparison between groups is seen as a way of observing the difference between *what happened when* **X** and *what would have happened when* **Not-X**.[13]

Depending on your background, all of this might seem obvious, and, in fact, nothing I've explained so far would be out of place in an Introduction to Psychology class. If it seems obvious, that's because the counterfactual or potential-outcomes analysis of causation has been thoroughly embedded in scientific practice. The computer scientist Judea Pearl, a founding editor of the *Journal of Causal Inference*, went so far as to call counterfactual reasoning the "cornerstone of scientific thought."[14] When we ask whether an intervention caused children to perform better in school, or whether a medication caused a reduction in symptoms, or whether an advertisement increased sales, we are typically asking, "What is the average difference these things made in the world?"

What Causes Are Not

Given that the counterfactual analysis of causation is endemic to every branch of medicine and the social sciences, it seems reasonable enough to apply this same understanding of causation to genetic causes. A genetic cause is something that, in David Lewis's words, "makes a difference, and the difference it makes must be a difference from what would have happened without it."

This bears repeating. To call a gene a cause—indeed, to call *anything* a cause—is to imply a comparison with some alternative reality where that cause did not happen (**X** versus **Not-X**). To say that a gene has an effect is to say that the gene makes a *difference*.[15]

But before we follow that train of thought any further, it will be helpful to lay out, in advance, what boundaries this counterfactual framework imposes on our understanding of causation. By considering these boundaries in the context of the Romanian orphanage study, which is a relatively straightforward example of inferring causation from a randomized experiment, we will be able to see more clearly how the word "cause" can be applied to thinking about genetic causes.

First, the conclusion that being assigned to foster care *causes* an increase in children's IQs does not imply that researchers know the *mechanism* for *how* this works. For example, one potential mechanistic story for the effects of foster care goes like this: Proximity to a warm and responsive caregiver downregulates physiological reactivity, thus preventing glucocorticoids from interfering with the development of synaptic connections necessary for learning and memory. Another one goes like this: Foster care families are more likely to feed children diets that have sufficient levels of iodine. Another one goes like this: The child brain is an "experience-expecting" organ, and without sufficient exposure to language in very early childhood, there is an insufficient proliferation of synapses in the child cortex.

Each of these mechanistic stories could be decomposed into a set of sub-mechanisms, a matryoshka doll of "How?"—this is how the brain encodes information about caregiver proximity, this is how glucocorticoids affect neurons in the forebrain, this is how the body metabolizes iodine, etc., etc. To say that we *understand* the effects of foster care on cognitive development requires working out these mechanisms.

But understanding mechanism is a *separable* set of scientific activities from those activities that establish causation. In ordinary scientific discourse, we are perfectly comfortable using the word *cause,* even when the mechanisms that instantiate that cause are almost entirely unknown. Being moved out of institutional care causes an increase in IQ, but how? *No one really knows.*

Second, identifying something as a cause in a counterfactual framework does not imply that cause *determines* the effect, only that the cause raises the *probability* that the effect will occur. In the course of customary scientific practice, and in the course of everyday life, we make claims about indeterministic causes all the time. And those claims are going beyond the statement that things are merely correlated: they are claims based on the outcomes of experiments where being randomly assigned to experience something raises the probability of one experiencing a particular outcome but does not determine that outcome. Psychotherapy combined with antidepressants raises the probability that depressed teenagers will stop thinking about suicide (but doesn't work for everyone).[16] Exercise lowers the probability of gaining weight (but some people still struggle to maintain their weight even when they exercise).[17] Taking enough folic acid when you're pregnant lowers the probability your baby will be born with a neural tube defect (but doesn't totally eliminate the possibility).[18] Suicidal thoughts and weight gain and neural tube defects are chancy events, but this doesn't stop us from using the language of cause and effect. We use the language of cause and effect to talk about our power to change people's chances.

Looking back at the IQs of Romanian orphans, the average IQ for those who remained in the orphanage was 73, but there was still variability. Some children who remained in orphanages nevertheless had higher IQs than those who were put into foster care. Let's define an arbitrary cut-off, in which we say that "normal" IQ is 70 or higher. (This cut-off, while arbitrary, is also meaningful: for one, it is the same cut-off that determines whether or not one can be executed for committing a crime in the United States.) Being assigned to foster care causes an increase in the *probability* that an individual child will develop "normal" cognitive abilities, but nothing is certain. Even as radical an intervention as changing literally every single thing about the child's environment—what he eats and where he sleeps and how he learns and which people take care of him and how lovingly and consistently they do it—is not enough to *determine* a certain level of cognitive ability. But this ordinary indeterminism does not disqualify something as a cause.

Third, in the absence of deterministic causation, one cannot make any confident claims about what caused the outcomes *for a particular individual*. Let's consider an individual child who was adopted from a Romanian orphanage and who had an IQ of 82. How much of that individual child's IQ was caused by being adopted? *We don't know*. We can say that, on average, children who were adopted had IQs that were 8 points higher than the IQs of children who were not. We *cannot* say, however, that one particular adopted child had a higher IQ than one particular institutionalized child because the former was adopted, or that 8 out of a child's 82 IQ points were due to their being adopted.

Finally, the *portability* of a cause can be limited or unknown. We can describe the results of the Romanian orphanage study as if it gives us insight on the benefits of foster care over institutional care, but are these results true of *all* orphanages or *all* foster care homes in all times and places? What if the study had been done in New Jersey in 2019? What if the study had been done in sixteenth-century France?

The developmental psychologist Urie Bronfenbrenner referred to the "bioecological" context of people's lives.[19] Everyone is embedded in concentric circles of context, each of which is mutually influencing the other. Closest to a person is her micro-context, comprising her immediate relationships and surroundings: families, friends, schools, neighborhoods, daily institutions. Who and what do you see and talk to every day? What air are you breathing, what water are you drinking, what food are you putting into your body? These micro-contexts exist within macro-contexts of political systems, economies, cultures, with various institutions (such as schools and workplaces) operating as intermediaries between the macro-system and one's day-to-day relationships.

I find Bronfenbrenner's bioecological model to be a helpful framework for thinking about the portability of causes of human behavior: Which of these circles would have to change, and by how much, in order for the causal claim to no longer be true? Here, knowing about mechanism also helps knowing about portability, as a good understanding of mechanism allows one to predict how cause-effect relationships will play out even in conditions that have never been observed. "The action of the sodium-potassium pump causes a neuron to have electrical potential" is a causal claim that is highly portable, regardless of what changes about Bronfenbrenner's circles. It is as true in ancient hunter-gatherers as it is in North Koreans in the early twenty-first century. "Adopting a child into foster care causes higher IQ"—this causal claim is likely less portable across different permutations of the bioecological context.

Precisely how portable genetic associations are, across time and place, is an empirical question that is just now beginning to be addressed with data. For example, researchers have found that polygenic indices are more strongly associated with educational attainment in more recent generations of women, who have greater access to educational opportunities, than in previous generations, who faced greater social obstacles to schooling.[20] We will come back to more examples like this in chapter 8. For now let us simply observe that limited portability is not, in and of itself, incompatible with causality.

An insistence on perfect portability as necessary for causation has animated some the most enduring criticisms of behavioral genetics. The evolutionary biologist Richard Lewontin, who was a vociferous critic of behavioral genetic studies of human behavior, alleged that scientific results that have a "historical (i.e., spatiotemporal) limitation" and that do not give information about "functional relations" (i.e., mechanisms) are "no use at all."[21]

In contrast to this stance, consider when scientists test whether cognitive-behavioral therapy reduces bulimic symptoms, or whether a public-school sexual health curriculum reduces syphilis rates, or whether iPhones increase teenage suicidality—all of which are certainly socially and historically specific phenomena. If one steps back from focusing on genes as causes, and instead uses a wide-angle lens to consider all the different types of causes that are typically studied by history, economics, sociology, political science, and psychology—that is, *all of social science*—the insistence that causes must have perfect portability begins to seem bizarre. It's far more useful to grade portability on a sliding scale, from "this only happens in the lab on rainy Tuesdays" to "this is a law of nature that we can expect to be true of humans at all times and places."

Thick and Thin Causation

In the course of ordinary social science and medicine, we are quite comfortable calling something a *cause,* even when (a) we don't understand the mechanisms by which the cause exerts its effects, (b) the cause is probabilistically but not deterministically associated with effects, and (c) the cause is of uncertain portability across time and space. "All" that is required to assert that you have identified a cause is to demonstrate evidence that the average outcome for a group of people *would have been different* if they had experienced **X** instead of **Not-X**. And the most convincing evidence that you know what might have been is to assign people *randomly* to **X** or **Not-X**. (The word "all" is in scare quotes here, because as any scientist of human behavior and society knows, actually isolating the variable of interest from the web of potential confounds, so that one can make an inference about causation, turns out to be an incredibly difficult and delicate operation.) I'm going to call this a "thin" model of causation.[22]

We can contrast the "thin" model of causation with the type of "thick" causation we see in monogenic genetic disorders or chromosomal abnormalities. Take Down's syndrome, for instance. Down's syndrome is *defined* by a single, deterministic, portable cause. To have three copies of chromosome 21, instead of two, is the necessary, sufficient, and sole cause of Down's syndrome. The causal relationship between having three copies of chromosome 21 and Down's is one-to-one, with the result that forward and reverse inferences work equally well. The cause of Down's is chromosome 21 trisomy; the effect of chromosome 21 trisomy is Down's. Having three copies of chromosome 21 doesn't raise your *probability* of having Down's; it is deterministic of the condition. And this causal relationship operates as a "law of nature," in the sense that we expect the trisomy-Down's relationship to operate more or less in the same way, regardless of the social milieu into which an individual is born.

Much of the furor about whether there are genetic causes of complex human outcomes, like educational attainment, is stoked by the fact that people—scientists and the lay public alike—think that genes always have to be "thick" causes. That is, people think that genes always have to operate like they do in Down's syndrome. As a social scientist, when I say that genes cause behavior, I'm making a probabilistic statement about a counterfactual—*if* your genes had been different, then there is a non-zero probability that your life would have been different. I am *not* claiming that any particular DNA sequence is a necessary or sufficient cause of one's life outcomes, that DNA determines anything about your life, that this counterfactual is perfectly portable across time and place, that I can retroactively infer that the capital-C Cause of your life is your genes, or that I even know how any stretch of DNA works.

Random Genes?

I hope I have convinced you that, if **X** versus **Not-X** is randomly assigned, then observing differences in outcomes that are probabilistically associated with **X** versus **Not-X** is satisfactory evidence that **X** is a "thin" cause of those outcomes. But geneticists are (currently) not conducting genetic experiments with humans, randomly assigning them to one genotype or another. So, how do we say anything about genes as *causes*? We will turn our attention to this question in the next chapter.

6

Random Assignment by Nature

In the Biblical book of Genesis, the origin story of the world is scarcely underway before we are confronted with a pair of brothers who chose different occupations: "Now Abel kept flocks, and Cain worked the soil. . . . The Lord looked with favor on Abel and his offering, but on Cain and his offering he did not look with favor" (Genesis 4:2–5). We all know how well *that* ended. Only one generation from creation, brothers were rewarded unequally for their labor, and the seething resentment provoked by that inequality led to humanity's first murder.

Why did these brothers end up with different lives—working the soil versus keeping flocks, committing violence versus being its victim? As I explained in the last chapter, if we were interested in testing the environmental causes of occupational choice or aggressive tendencies, we might decide to run an experiment. Let's say we randomly assign one group of families to have access to a high-quality preschool, whereas a comparison group is left to their own devices. Do the children in the first group grow up to make different choices in the labor market? Do they grow up less likely to commit violent crimes? If we do our experiment correctly, we will still not know the mechanism by which preschool experiences affect aggression. We will not know how portable that effect is across sociopolitical and historical contexts. But we will be confident in saying whether or not high-quality preschool, *on average,* caused a decrease in violence.

What about genetic causes instead of environmental ones? It is essentially impossible (for now), and certainly unethical, to run an experiment where we randomly select a group of children and edit their genomes *in utero,* in order to test whether those genetic changes have a causal effect on their life outcomes. Fortunately for science, however, we don't need to run an experiment to randomly assign children to genes, because nature is already running that experiment for us.

Remember that humans have two copies of every gene, but only one of these copies is passed down from a parent to a child. Every time a child is conceived, then, which of his parents' two genes he inherits is randomly assigned. In this way, genetic inheritance is working just like the Romanian orphanage study that I described in the last chapter. In the Romanian orphanage study, children who went to foster care differed from children who stayed in the orphanage because of luck—an experimenter picked their name out of a hat. In the course of everyday life, children who inherited genetic variant **X** from their parents differ from their siblings who did *not* inherit variant **X** because of luck. Instead of experimenters being the arbiters of luck, however, it was nature itself.

Given a set of parental genotypes, which genes their children inherit is random. Consequently, comparing siblings who differ genetically helps you draw causal conclusions about the average effects of genes for people's life outcomes. If one group of people all inherited variant **X**, and *their siblings* all inherited variant **Not-X**, then comparing the educational attainment of those two groups allows you estimate the average causal effect of variant **X** on educational attainment. The logic of the sibling comparison is the exact same logic as any randomized controlled trial of a medication or any experimental study of an environmental intervention.

Put differently, if siblings who differ genetically also have corresponding differences in their health or well-being or education, this is evidence that genes are *causing* these social inequalities. We can use the natural lottery of sibling differences in genetics to examine whether genes influence how adroitly people manipulate abstract information, how organized or impulsive they are, how far they go in school, how much money they make, how happy and satisfied they are with their lives. (In saying this, remember the caveats of "thin causation" from the last chapter: I am *not* claiming that any particular DNA sequence is a necessary or sufficient cause of one's life outcomes, that DNA determines anything about your life, that genetic causes work the same across all social and historical contexts, that I can retroactively infer that the capital-C Cause of your life is your genes, or that I know how any stretch of DNA works.)

In this chapter, then, I'll describe research that has compared siblings, or other types of biological relatives, in order to test whether the genetic lottery causes differences in life outcomes.

Each Unhappy Family Member Is Unhappy in His Own Way

My brother Micah was born three years after me. We are unmistakably siblings. We have the same brown hair, same green eyes, same tendency to do what our stepmother refers to as the "Harden slow-blink," closing our eyes for a few seconds when we are annoyed at someone. Sometimes he sends me *R* code for functions he's written, and I feel loved.

Despite these similarities, our lives have turned out differently. I have six more years of formal schooling, have never been unemployed, make more money, have given birth to two children. He is still married, lives close to our family and childhood friends, is blissfully free of the neuroticism and ADHD symptoms that plague my daily life, can still run up and down a soccer field without gasping for oxygen.

For each life outcome, we might ask how much the randomly occurring genetic differences between us nudged us down diverging life paths. But let's start with an easy one—height. Micah is 5′9″ (1.75 meters), shorter than the average American man, whereas I am 5′7″ (1.70 meters), taller than the average American woman. Did we end up with different heights because we inherited different genes? The answer to that might seem obvious, but it's worth going through how scientists might approach answering that question.

Many scientists' first response to that question—do Micah and I differ in our heights because we inherited different genes—would be to point out that it's the wrong question. As I described in the previous chapter, outcomes like height (not to mention more complicated social outcomes like education) are influenced by many genes, which are *probabilistically* related to the phenotype. We typically observe probabilities by studying the frequency of certain outcomes in *groups* of people who lived in a particular time and place. As a result, we typically cannot say anything about whether genes have caused something for an individual person's life. Such inferences might sometimes be appropriate in the case of extremes, such as Shawn Bradley, the extraordinarily tall NBA player that I told you about in chapter 2. But, generally, the research designs that scientists use to connect genes to complicated human phenotypes allow us to test whether genes caused differences in height, on average, not whether one specific person is taller than another one because of her genes.

So let's reframe the question slightly: Do genes cause differences in people's height, on average?

One way to test this is by examining something called "identity by descent." When my mother's body was making the egg that became me, *her* paternal and maternal chromosomes swapped chunks of genetic material. So the chromosomes that I inherited from her have a 100 percent unique sequence of DNA, made up of alternating segments that can be traced back to either my maternal grandmother or my maternal grandfather. The same process played out when my mother's body was making the egg that became my brother Micah. And this whole process is further doubled because we also have a father.

If, then, you look at any one spot on one of the chromosomes we inherited from our mother, there is a 50/50 chance that Micah inherited the exact same DNA segment as I did. If he did, then we are what is called "identical-by-descent" (IBD) on that segment. Because we have two parents, and thus two copies of every chromosome, we could be essentially clones of each other for any one DNA segment,

Shared DNA
44.6%
3321 cM

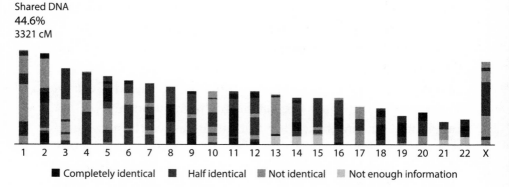

FIGURE 6.1. Identity-by-descent sharing of segments of 23 chromosomes between a pair of full siblings. Image from author's 23andMe® profile. The author and her brother share segments of DNA that have a total length of 3321 centimorgans (cMs), which is 44.6% of the author's genome.

i.e., share identity-by-descent for both of the segments inherited from our father and from our mother. Or, we could be essentially unrelated, with identity-by-descent sharing from neither parent. Or, we could be matching on a segment from one of our parents but not the other.

My brother agreed to be genotyped by 23andMe for the sake of this book. (In true little-brother fashion, he immediately demanded that I Venmo him $200 to pay him back.) Helpfully, 23andMe automatically generates an infographic showing which DNA segments we share, and which ones we don't (figure 6.1). On chromosome 11, for instance, we're nearly twins; on chromosome 13, we are barely related.

On average, we are expected to share 50 percent of DNA segments. But that's *on average*. If you flip a coin 1,000 times, the expectation is that it will land on heads 50 percent of the time, or 500 times. But in reality, it might land on heads 501 times. Or, even weirder, 545 times. Like flipping a coin, reproduction is a stochastic process. Two siblings are expected to share 50 percent of their DNA segments, but *in reality*, they might share a little bit more or a little bit less. Micah and I share a little bit less than the expectation—44.6 percent.

In 2006, the statistical geneticist Peter Visscher and his colleagues conducted a study that took advantage of the fact that there is random variation in the extent of identity-by-descent sharing between siblings—sometimes it's lower than 50 percent, sometimes it's higher.[1] For each pair of siblings, they divided the genome into segments called *centimorgans* (abbreviated cM) and calculated the actual number of 1-cM segments that were shared between the siblings. (Micah and I share 3,321 cM.) On average, siblings in this sample shared 49.8 percent of their DNA segments, which is remarkably similar to the theoretical expectation of 50 percent. But any individual pair could share more or less: the *range* of identity-by-descent sharing was 37 percent to 62 percent.

Next, Visscher and colleagues asked whether siblings who inherited more-*different* genotypes also showed greater dissimilarity in height. As I described in the previous chapter, this question—whether genes make a *difference* to one's height—is fundamentally a causal question. And the answer, perhaps not surprisingly, was yes.

Of course, siblings differ in their height for reasons other than genetics, too. (My brother refused to eat anything but Rice Krispies for most of 1989, which surely stunted his growth.) Siblings who inherited more-different genes might be different in their height, but those genetically caused differences might be a drop in the bucket compared to the other factors that make them taller or shorter. The relative effect of genes, then, can be expressed as a ratio: how different siblings are in their height because they inherited different genes, divided by how different people are in their height generally. This ratio has a name that might be familiar: it is the *heritability* of height. In this study of height, the researchers concluded that the heritability of height was about 80 percent. That is, about 80 percent of the total variance in height was due to the fact that people inherited different genes.

Heritability Is about Differences

Here, it will be useful to recap the argument I've been building over the last two chapters, because—like a frog being slowly boiled alive—we've gone from an uncontroversial premise to a highly controversial one. Beginning in the previous chapter, I started with the uncontroversial premise: comparing the average outcomes of two groups of subjects who have been randomly assigned to **X** or **Not-X** is a test of the average causal effect of **X**. Then, I pointed out that, conditional on their parents' genes, siblings are randomly assigned to genetic variants. Therefore, a comparison of siblings who differ genetically is a test of the causal effect of that variant—nature's truest natural experiment. This causal test can be described in terms of counterfactual dependence: *If one's genotype were different, then would one's life outcomes be different?*

In this chapter, I have begun to explain how researchers have gone about actually conducting this causal test. The Visscher study on height took advantage of between-sibling differences in genotype: If one's genotype is *more* different from one's sibling's, then is one's height more different from one's sibling's height? The results of this causal test, in turn, can be expressed in terms of a statistic that might be familiar, or at least familiar-sounding: the *heritability coefficient.*

In this way, we have arrived at a conclusion that is sure to provoke disagreement from some readers: heritability estimates, which quantify the extent to which differences in life outcomes are due to differences in genotype, are a test of whether genes have a *causal* effect on life outcomes.

Perhaps no concept in genetics has been the subject of as much confusion as heritability, which is a technical term that, unfortunately, sounds like an ordinary English word. The linguistic roots of the word "heritability" predate any knowledge about DNA by millennia. *Heres* was the Latin word for "heir," the (male) person who was legally entitled to someone's property and social rank upon their death. A "hereditary" aristocracy is a society in which wealth, ranks, titles, powers, and privileges are replicated from generation to generation. An "inheritance" is an asset that is transferred from parent to child. When we hear the word "heritability," it is almost impossible not to burden the word with several thousand years of cultural baggage about how "inheritance" works. Inheritance is about faithfully reproducing social hierarchies; inheritance is about unbroken continuity from parent to child.

But, as I discussed in chapter 2, humans do not "breed true." It is a mistake to imagine that heritable traits are those that are inherited, intact, from parent to child, because this conception ignores the fact that half of the genetic variation exists *within* families. I have two copies of every gene, and this internal genetic diversity is made manifest in genetic differences between my children.

To continue with the example of height, a heritability of 80 percent means that most of the differences in height *within the population being studied* (an important point to which I'll return at the end of this chapter) were caused by genetic differences between people. But these height-causing genetic differences exist *within* families as well as between families. If the standard deviation of adult male height in the population was 3 inches, with a mean of 70 inches, we would expect the distribution of heights in the population to look like the top of figure 6.2. Compare that distribution to the distribution of heights we would expect for all the potential male offspring of a father who is slightly above average in height (71 inches)—the bottom half of figure 6.2. The range of potential outcomes is narrowed somewhat—the children of slightly taller parents are less likely to be very short—but is certainly not eliminated.

Observing high heritability, then, does *not* mean that inequalities between people will be perfectly replicated across generations: Tall parents can sometimes have shorter children. In fact, high heritability implies that children of the same parents will diverge in their life outcomes. Heritability is about whether genetically different people show phenotypic differences, and siblings are genetically different.

The Heritability of Seven Domains of Inequality

We have discussed how sibling differences in identity-by-descent sharing could be used to estimate the heritability of height. This approach relies on measuring people's DNA, but the concept of heritability predates the technology to measure DNA. Throughout the past century, and even today, the most common method for estimating heritability has been to compare identical twins to fraternal twins.

News reports about twins or triplets often focus on babies who were separated at birth and grew up in different homes,[2] like the brothers from the movie *Three Identical Strangers*,[3] but the vast preponderance of twin studies have been conducted with twins who

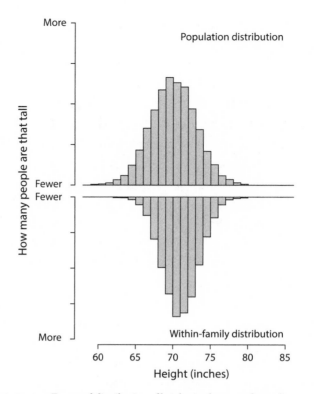

FIGURE 6.2. Expected distribution of heights in the general population (top) versus within potential offspring of a single pair of parents
(bottom). Population distribution is based on mean of 70 inches with
a standard deviation of 3 inches. Within-family distribution, i.e., the
distribution of heights among all possible offspring of a single pair of
parents, based on heritability of 0.8. Example and calculations adapted
from Peter M. Visscher, William G. Hill, and Naomi R. Wray, "Heritability in the Genomics Era—Concepts and Misconceptions," *Nature Reviews
Genetics* 9, no. 4 (April 2008): 255–66, https://doi.org/10.1038/nrg2322.

were all raised in the same home by their birth parent(s). From here
on out, unless I specifically refer to twins "reared apart," I am talking
about twins who are raised together in the same home. The basic logic
of this type of twin study is probably familiar. Consider pairs of identical twins—the Weasley twins from Harry Potter; the Winklevii (Cam-
eron and Tyler Winklevoss), who challenged Mark Zuckerberg's
claim to Facebook. Each pair began life as a single zygote, but a fluke
in cell division during the early stages of development resulted in

the formation of two zygotes from one. Identical, or monozygotic, twins are not 100 percent genetically identical 100 percent of the time, because, well, stuff happens. Genes mutate early in development but after the zygotes have split, resulting in genetic differences between twins, or even genetic differences between different parts of the body within the same person. And identical twins can—and do—also differ in their gene *expression*, which is whether and when the genes that they do have are turned "on" or "off" in different parts of their body.

But even with these differences, identical twins have been the focus of fascinated adoration and grisly curiosity throughout history, and they continue to be one of nature's most intriguing natural experiments: What happens when someone else begins life in the *exact* same place as you, indeed *is* you for their first few fleeting hours?

Fraternal twins, on the other hand, have a somewhat more prosaic beginning. They are just like non-twin siblings, each formed from their own unique combination of sperm and egg. The only difference is that those eggs were released during the same menstrual cycle, resulting in two fetuses in a single pregnancy.

All twins who were not "separated at birth" and adopted away into different homes, whether identical or fraternal, share everything about their initial social position, particularly as defined by most of the key variables of social science, such as Zip code and family income and school district. We'd expect, then, that twins will grow up to be similar to one another. Key to a twin study is the question: How much *more* similar are identical twins than fraternal twins?

All pairs of twins raised in the same home are exposed to the same set of parental foibles, to the same neighborhood conditions, to the same schools. But identical twins share *more*. They also share (nearly) all of their genetic code. Or, put differently, fraternal twins share *less*. They are more different from each other, genetically, than identical twins. Just as with the sibling study on height that I described at the beginning of this chapter, the key question for testing the causal influence of genes on life outcomes is this: Are people who are more genetically different (in this case, fraternal twins compared to identical twins) also more phenotypically different? The

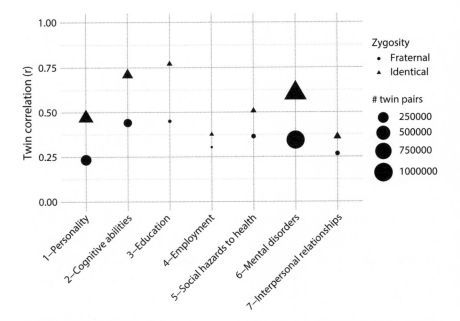

FIGURE 6.3. Identical and fraternal twin correlations for seven domains of inequality. Author's analysis of data from Tinca J. C. Polderman et al., "Meta-Analysis of the Heritability of Human Traits Based on Fifty Years of Twin Studies," *Nature Genetics* 47, no. 7 (July 2015): 702–9, https://doi.org/10.1038/ng.3285.

more different fraternal twins are in a particular trait, like height, in comparison to identical twins, the higher the heritability of that trait.

In 2015, a paper in the journal *Nature Genetics* summarized fifty years of twin research—over 2,000 scientific papers on over 17,000 traits measured in over 2 million twin pairs.[4] From their paper, I pulled data on seven different life domains, which I've plotted in figure 6.3.

The first two domains are aspects of a person's psychology—personality characteristics and cognitive abilities. These psychological traits are important because they are the psychological traits that are most strongly correlated with the third outcome: success in education.[5] Education, in turn, is a strong factor in determining one's success in the fourth domain: the labor market. People who are unemployed and/or have low incomes experience difficulties in the fifth domain—social hazards to health, like living in a poor neighborhood where rates of pollution and violence are higher. The last two domains are risk for mental disorders, such as depression or alcoholism, and interpersonal relationships, such as whether or not one is married or divorced, reports feeling lonely, sees friends, etc.

As you'll see from the size of the bubbles in figure 6.3, there has been a mountain of twin research, with over 1 million twins contributing data to twin studies of mental disorders. And as you'll see from the gaps between the circles and the triangles, in every domain, fraternal twins are more different than identical twins. The greater the distance between the fraternal twin correlation and the identical twin correlation, the greater the heritability.

What this graph is showing us is that all seven of these domains of inequality—cognitive ability, personality, education, employment, social hazards to health, mental disorders, and interpersonal relationships—are substantially heritable, with about one-quarter to one-half of the variation due to differences in inherited DNA sequence. After fifty years and more than 1 million twins, the overwhelming conclusion is that *when people inherit different genes, their lives turn out differently.*

A Familiar Objection

Even as I write this, I hear a chorus singing out a familiar objection: "Heritability estimates are specific to a population." That is, even if these domains of inequality are heritable among the specific groups of people being measured, these heritabilities are not fixed laws of nature that are true across all times and places and groups of people. Like GWAS research, twin research has a clear Eurocentric bias in who has been studied: White adults living in Minnesota and Colorado and Texas and Wisconsin and Virginia and the Netherlands and Norway and Denmark and Finland and Sweden and the UK and Australia during the twentieth and early twenty-first centuries. If different people, living in a different time and place, and experiencing different social structures, had been studied, or were studied in the future, then heritabilities of life outcomes might also be different. To pick just one example, as a woman I would not have been allowed to attend college, much less complete a doctoral degree, if I had been born in 1782 instead of 1982. As the environmental opportunities for attaining education changed, so too did the relevance of whatever genes I happened to inherit. In the coming chapters, I will discuss multiple empirical examples like this, showing how heritability differs across social and historical contexts.

The fact that heritabilities can (and do) differ across populations has been a major sticking point for critics of behavioral genetics, who have advocated that the concept, and its estimation, be abandoned entirely. Here is the biologist Richard Lewontin in 1974: "I suggest we stop the endless search for better methods of estimating useless quantities."[6] Here is the psychologist Richard Lerner in 2004, bemoaning, "Why do we have to keep reinterring behavior genetics?"[7] Here is the economist Charles Manski in 2011 asking, "Why does heritability research persist? . . . The work goes on, but I do not know why."[8]

But we don't treat other population statistics about inequality as unimportant because of their specificity to a particular time and place. The Gini index, for instance, is a measure of income inequality. A country where everyone makes the exact same income has a Gini index of 0; a country where one person makes all the money and everyone else has nothing has a Gini index of 1. Just as a trait doesn't have a single heritability, a country doesn't have a single Gini index—it changes over time with economic and political changes. If someone used the Gini to describe the inequality that people in a particular society experienced at a particular moment in historical time, we would not rush to dismiss the information as "only" population specific.

Heritabilities and polygenic index associations are like the Gini index—historically and geographically specific, yes, but no less interesting or valuable in being so.[9] Even if heritability estimates are entirely population specific, they remain an important summary of how much inequalities in life outcomes were caused by the outcome of the genetic lottery *for that population*. Despite pleas to abandon the concept, I anticipate that heritability research will persist, for good reason. Heritability research persists because it is answering a question about whether people's genes, an accident of birth over which they have no control, caused differences between people in things we care about—differences in education and income and well-being and health—in the societies in which we actually live.[10]

The Case of the Missing Heritability

Especially in the wake of Herrnstein and Murray's *The Bell Curve*, the assumptions of twin studies have come under close scrutiny. And, politically-motivated or no, there are good reasons to scrutinize the assumptions of twin studies. They *do* indeed make a lot of assumptions, many of which might not strike you as particularly plausible. For one, the twin study assumes that identical twins aren't treated more similarly to one another *just* because they are identical—the "equal environments assumption." If you've ever seen twins dressed in outfits that perfectly match, down to their socks and hair bows, that assumption might seem like a bit of a stretch.[11] More generally, genes and environments are correlated in complicated ways that can be difficult to measure and statistically account for, leading to the persistent suspicion that maybe twin studies are attributing to genes what should really be claimed by the environment.

The results of early GWAS fed the suspicion that twin studies were getting something fundamentally wrong. As I described in chapter 3, the top "hits" from a GWAS of educational attainment in over 1 million people were worth, at most, just a few weeks of additional schooling, accounting for just a fraction of 1 percent of the variation between people in how far they went in school.[12] If you put all the genes identified in the GWAS together in the form of a polygenic index, you can account for ~13 percent of the variance in educational attainment. This is substantial when viewed in comparison to the effect sizes that we see for other social science variables (e.g., 11 percent for family income).[13] Yet it is still a far cry from the twin study estimate that about 40 percent of the variation in educational attainment is due to genes.[14]

This gap between the variance accounted for by genes discovered in GWAS and the heritability estimated from twin studies has been called the "missing heritability" problem (figure 6.4).

FIGURE 6.4. The case of the missing heritability. Image reproduced by permission of Springer Nature from Brendan Maher, "Personal Genomes: The Case of the Missing Heritability," *Nature* 456, no. 7218 (November 1, 2008): 18–21, https://doi.org/10.1038/456018a.

But before we use the phenomenon of "missing heritability" to dismiss the conclusions of twin studies out of hand, it's important to remember that there are also reasons to suspect that GWAS and polygenic index studies likely *under*estimate the effects of genes, for at least two reasons.[15] First, these methods don't measure every genetic variant, particularly rare genetic variants which might have especially large effects. And, second, even GWAS of over 1 million people might still not have enough people to detect very weak, but still non-zero, effects of individual genes.

If heritability estimates from twin studies might be too high, but the estimates of genetic effects from GWAS might be too low, what then is the best estimate for the impact of inherited DNA variation on life outcomes such as education? Ultimately, as the statistical geneticist Alex Young explained, "the deepest solution to the missing heritability problem would involve identifying all of the causal genetic variants and measuring how much trait variation they explain."[16]

We are, obviously, not there yet for any human phenotype, much less complicated ones like education. In the meantime, one method of obtaining a Goldilocks (not too big, not too small) estimate of heritability is the sibling regression method that I told you about in the beginning of this chapter, which uses random variation among sibling pairs in extent of identity-by-descent sharing. This method can be extended to use other types of biological relatives, in a method called relatedness disequilibrium regression, or RDR.[17]

In figure 6.5, I've plotted heritability estimates obtained from using sib regression, RDR, and twin methods for four outcomes: (1) height, (2) BMI, (3) age at first birth in women, and (4) educational attainment. For education, there remains some uncertainty whether genes cause 40 percent of the variation in people's outcomes, or something closer to 17 percent. But by comparison, recall that family income accounts for just 11 percent of the variation in educational attainment among White-identifying people in the United States.[18] What this comparison shows us is that even when you throw out the controversial assumptions of twin studies, the heritability of educational attainment is *still* not zero. Genes cause differences in educational outcomes, and at a minimum, the effects of those genetic differences are at least as important in explaining variability as a variable like family income.

Within-Family Studies of Polygenic Indices

Heritability studies using twins or siblings are telling you something about the overall effect of the entire genome on people's life outcomes, but they don't tell you about which specific genetic variants are driving that effect. In contrast, GWAS aims to identify specific genetic variants, but typical GWAS studies compare people from different families and so are always in danger of picking up on environmental effects that just happen to be correlated with genetic

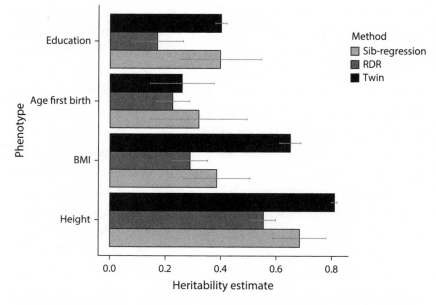

FIGURE 6.5. Heritability estimates for four human phenotypes from three different methods. "Education" = educational attainment (years of formal schooling). "Age first birth" = women's age at first childbirth. "BMI" = body mass index. "Height" = height in adulthood. "Twin" method estimates heritability by comparing similarity of monozygotic twins reared together to similarity of dizygotic twins reared together. "Sib-regression" method estimates heritability by leveraging random variation among sibling pairs in extent of identity-by-descent sharing. "RDR" (relatedness disequilibrium regression) method extends the sib-regression method to other pairs of relatives, where the relatedness of the pair is conditioned on the relatedness of their parents. Error bars represent standard errors. All heritability estimates drawn from Alexander I. Young et al., "Relatedness Disequilibrium Regression Estimates Heritability without Environmental Bias," *Nature Genetics* 50, no. 9 (September 2018): 1304–10, https://doi.org/10 .1038/s41588-018-0178-9, except for twin estimate of heritability for educational attainment, which is drawn from Amelia R. Branigan, Kenneth J. McCallum, and Jeremy Freese, "Variation in the Heritability of Educational Attainment: An International Meta-Analysis," *Social Forces* 92, no. 1 (2013): 109–140; and twin estimate of heritability for age at first birth in women, which is drawn from Felix C. Tropf et al., "Genetic Influence on Age at First Birth of Female Twins Born in the UK, 1919–68," *Population Studies* 69, no. 2 (May 4, 2015): 129–45, https:// doi.org/10.1080/00324728.2015.1056823.

differences. A way of merging these two approaches is to use GWAS results to construct a polygenic index, and then test the polygenic index using a sample of family members. When we are looking at how the genetic lottery plays out in a single generation, genetic differences between family members are random, rather than braided together with between-family differences in ancestry and geography and culture. Within-family approaches, then, capitalize on nature's experiment to test whether the specific genes captured by a polygenic index cause differences in life outcomes.

There are three different types of within-family studies that researchers have used to investigate the effects of the genetic lottery (1) sibling comparison studies, (2) studies comparing adoptees and non-adoptees, and (3) studies of parent-offspring trios.

The sibling design is perhaps the most straightforward: Do siblings who differ in their polygenic index differ in their life outcomes? One study in this vein followed over 2,000 pairs of fraternal twins in the UK from the time they were 12 until they were 21.[19] The researchers measured each person's height, body mass index (BMI), self-rated health, ADHD symptoms, psychotic experiences, neuroticism, intelligence test scores, and academic achievement as measured by scores on the GCSE (General Certificate of Secondary Education, a standardized test, akin to the SAT in the US, taken around age 16). For each life outcome, researchers could test whether siblings who were more different in their polygenic index showed more differences in their life outcomes.

They did. The twins who differed the most genetically differed in their actual height by nearly 9 cm (~3.5 inches). They differed in their BMI by 3 points, which is equivalent to gaining 20 pounds on a 5'7" woman. They differed in their GCSE scores by 0.5 standard deviations.

This UK twin study followed twins until they were twenty-one, but of course, there is lots of life left to live when you are twenty-one. How do people fare as they begin to be "real" adults, with marriages and mortgages?

This question was addressed in a study that I briefly told you about in chapter 2, by Dan Belsky, a sociogenomics researcher at Columbia University, and his colleagues.[20] Using a polygenic index based on the educational attainment GWAS, Belsky and his colleagues found that siblings who had higher polygenic indices *than their co-siblings* went further in their education, were employed in more-prestigious occupations, and were wealthier at the end of their working lives. As these sibling differences in genetic variants are the outcome of an entirely random Mendelian lottery, this study provides some of the most compelling evidence that one's genetics *cause* differences in education and wealth. (Of course, like many causal inferences from randomized controlled trials in the social sciences, we still don't know *how* these genetic effects are operating, just that they are—a topic we will come back to in the next chapter.)

The idea of using adoptees to study the genetic lottery was showcased in an ingenious study using data from the UK Biobank.[21] Adoptees, of course, are not raised by their biological parents. This means that their genetics are not related to their parents' genetics—and so their own genetics are less bound up with the complicated web of ancestry, geography, social position, and culture that is correlated with their parents' genetics. In a study of over 6,000 people in the UK, Cheesman and colleagues showed that a polygenic index was indeed associated with educational attainment in adoptees, but the strength of the relationship was weaker than in non-adoptees. This study thus provides evidence for the "direct" effect of one's own genes on educational attainment, while also raising the question of why the polygenic index is more strongly related to outcomes in children raised by their biological parents (a question that we will return to in chapter 9).

The final type of within-family design measures DNA in parent-offspring trios (two biological parents and their child). For each parent, their genome can be divided into two parts—the genes that have been transmitted to the child and the genes that were untransmitted. Again, which genes are transmitted versus untransmitted for each child is *random,* the outcome of the genetic lottery. The extent to which the transmitted genes are more strongly related to a child's outcome than the untransmitted genes, then, is a test of the causal effect of genes. Both the transmitted and untransmitted genes of a parent are correlated with aspects of the parent's ancestry, environment, geography, and culture, but only the transmitted genes were (randomly) inherited biologically.

The most high-profile study using this method was conducted in Iceland.[22] Studying over 20,000 people who had been genotyped, and their parents, the researchers drew a conclusion similar to that from the sibling comparison studies and the adoption study. The association between a polygenic index and educational attainment is attenuated when you compare within families—but it certainly doesn't go away. Results from all three methods, then, triangulate on the same conclusion: that the outcome of the genetic lottery has a causal effect on how far one goes in school.

Zooming back out, when we put together results from fifty years of twin research with results from just a few years of research using measured DNA, the inescapable conclusion is that genetic differences between people cause social inequalities—including inequalities in educational attainment, but also in physical health outcomes such as BMI, psychological outcomes like ADHD and other mental disorders, and fertility outcomes like age at first birth.

In 1962, the evolutionary biologist Theodosius Dobzhansky[23] wrote that "people vary in ability, energy, health, character, and other socially important traits, and there is good, though not absolutely conclusive, evidence that the variance of all these traits is in part genetically conditioned. Conditioned, mind you, not fixed or predestined." Dobzhansky was right, and the evidence that has accumulated in the decades since his writing has only served to make the case more conclusive.

It is unfortunate that so much energy has been wasted debating this fact, which was evident to Dobzhansky and others a half century ago, because determining that genes are a cause of social inequality is perhaps the easiest part of the research enterprise. A much more difficult question, to which we will turn our attention in the next chapter, is: How?

7

The Mystery of How

In 1998, I won a merit scholarship to Furman University, a small, formerly Baptist liberal arts college in South Carolina. The scholarship covered 100 percent of my tuition and fees for four years, including a study-abroad semester in London. Current tuition at Furman is nearly $50,000 per year. I paid nothing. Then, as now, I was entirely non-athletic. My extracurriculars were uninspiring. I had overcome no particular hardship, showed no particular resilience in the face of challenge or adversity. My sole form of so-called merit was a nearly perfect score on the Scholastic Aptitude Test (SAT), the college admissions test that is a rite of passage for American high school students.

As I described in chapter 6, we can be fairly confident that genes have a causal impact on one's educational attainment. And, as I described in chapter 5, statements about causality are not statements about mechanism. Just as we don't know much about how rescuing Romanian orphans increases their IQ test scores, we also don't know much about how one's genes ultimately affect success in education.

But we do know some things about how educational systems work. Consider my own journey to college. My parents, both of them college-educated, had clear expectations that I would also attend college, and some knowledge about how to navigate the admissions process. I also had access to other forms of social capital,

like a friend who had attended Furman the previous year and a high
school guidance counselor who suggested possible schools. At the
institutional level, Furman—like many small liberal arts colleges—
uses "merit" scholarships to attract students with high test scores.
This improves their ranking in the *U.S. News* list of "best" colleges,
and higher rankings make the school look better to other potential
students, the ones who pay full tuition.

When people imagine possible mechanisms for "genetic" effects
on complicated human outcomes, like education, social processes
like these—parental expectations, access to advantaged social net-
works, institutional jostling in a commodified educational market—
aren't the first things that spring to mind. Instead, it is easy to jump
to the conclusion that genetic causes must have entirely biological
mechanisms, happening inside the skin.

But answering the question of "how" involves studying not just
interactions between molecules and cells, but also interactions
between people and social institutions. In this chapter, I aim to
describe what we do know—or at least think we know—about the
question of how. In particular, I'll focus on pathways between the
genome and educational success, both because that is an area my
own research group has done extensive work in, and again because
of the centrality of education in structuring other inequalities.

As I described in chapter 4, it's important to remember that
everything I'm describing here applies to understanding individual
differences within groups. The research tools that I'm describing
here (primarily twin studies and polygenic index analyses) can't tell
us anything about the causes of average differences between groups.

Red-Headed Children and Alternative Possible Worlds

In 1972, the sociologist Sandy Jencks proposed one of the most enduring thought experiments about social mechanisms for genetic effects:[1]

> If, for example, a nation refuses to send children with red hair to
> school, the genes that cause red hair can be said to lower read-
> ing scores. . . . Attributing redheads' illiteracy to their genes

would probably strike most readers as absurd under these circumstances. Yet that is precisely what traditional methods of estimating heritability do.

Jencks was right. Estimates of heritability *do* provide information on whether genes cause a phenotype. But these designs *don't* provide information about what mechanisms connect genotypes and phenotypes, and the relevant mechanisms might not be intuitively "biological" processes.

Jencks's thought experiment about red-headed children has become a meme in discussions about genetics. It's an enduring idea, I think, because it intuitively captures three ideas that are worth unpacking in more detail—(1) causal chains, (2) levels of analysis, and (3) alternative possible worlds.

First, the red-headed child example makes clear that genes can be connected to phenotypes via long *causal chains*.[2] In this example, a variant of the *MC1R* gene codes for pheomelanin, which makes one's hair visibly red. This phenotypic characteristic is then perceived by others in terms of culturally and historically specific social biases, and these biases are entrenched in social policies that forbid certain children from going to school.

Second, these causal chains can span multiple *levels of analysis.* One way of organizing scientific inquiry is to arrange the phenomena that are investigated by scientists like a layer cake, where the objects in each layer are parts of the objects in the next layer up.[3] Subatomic particles, such as quarks, are parts of atoms; individual people are parts of societies (figure 7.1). Jencks's thought experiment about red-headed children makes clear that causal chains can stretch up through multiple levels of analysis: The *MC1R* gene is part of the DNA molecule. It creates pheomelanin, a protein, within cells. "Red-haired" describes an individual person, and the decision to forbid redheads from attending school is a social phenomenon.

When Jencks asserts that "attributing redheads' illiteracy to their genes would probably strike most readers as absurd," he is making an argument, in part, about the *best* level of analysis for describing and understanding the phenomenon: it is, in his view, absurd to

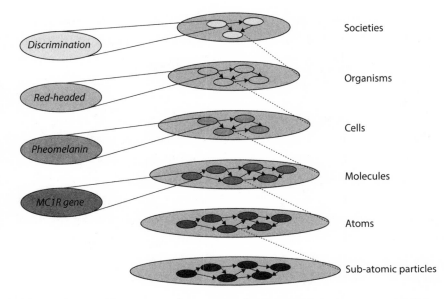

Societies

Discrimination

Organisms

Red-headed

Cells

Pheomelanin

Molecules

MC1R gene

Atoms

Sub-atomic particles

FIGURE 7.1. Levels of scientific analysis. Figure incorporates ideas from Carl F. Craver, *Explaining the Brain: Mechanisms and the Mosaic Unity of Neuroscience* (Oxford: Oxford University Press, 2007); Paul Oppenheim and Hilary Putnam, "Unity of Science as a Working Hypothesis," 1958, http://conservancy.umn.edu/handle/11299/184622; and Christopher Jencks et al., *Inequality: A Reassessment of the Effect of Family and Schooling in America* (New York: Basic Books, 1972).

consider going to school a molecular phenomenon, even if part of the causal chain involves a DNA molecule.[4]

And, third, in Jencks's example it is readily apparent that there are *alternative possible worlds* in which the causal chain from genes to illiteracy is broken. When you hear the red-headed child example, you can immediately imagine an alternative society where all children are allowed to go to school regardless of their hair color. That change in social policy would break the causal chain between genotype and phenotype without directly manipulating anything about children's genes or gene products. One does not have to edit embryonic DNA or give children pharmaceuticals to change their biology in order to ameliorate a "genetic" effect on education.

We can contrast red-headed children with, say, the relationship between *HTT* and Huntington's disease. First, the causal chain between *HTT* and Huntington's disease is relatively short, i.e., it doesn't require that many steps to explain the sequence from beginning to end. Second, the causal chain between *HTT* and Huntington's all takes place at levels of analysis that are obviously "biological." That is, to describe the causal chain, you describe actions of molecules within cells. And, third, it is difficult to imagine an alternative possible world in which *HTT* does not cause Huntington's, and the possibilities that one *can* imagine for breaking the *HTT*-to-Huntington's chain involve directly manipulating some aspect of the individual's biology, e.g., gene editing or pharmacology.

We can also envision causal chains that have neither the strict biodeterminism of Huntington's nor the pure social dependency of Jencks' hypothetical redheads. For example, one can intervene to improve the academic function of children with ADHD by using stimulant medication (changing molecules in cells), using behavioral strategies that provide prompt rewards for the completion of required tasks (changing the behavior of individuals), *and* by changing how classrooms are structured (changing social organizations). ADHD is neither purely biological nor purely social; it is a pattern of experience and behavior that arises at the intersection of someone's particular neurobiology with the expectations of a particular social context.

A heritability analysis or within-family polygenic index analysis doesn't tell you anything, *on its own*, about whether the genes causing lower educational attainment are operating more like "redheads-can't-go-to-school" genes or more like *HTT* genes. Even if we know that genes have a causal effect on social inequality, there remain important questions about the mechanism of those effects: How long is the causal chain, what are its links, how many levels of analysis does it span, and—perhaps most crucially for debates about social policy—how can the causal chain best be broken, or strengthened?

Beginning with *Hereditary Genius*,[5] the nineteenth-century book by the father of eugenics, Francis Galton, and continuing through the twenty-first century, with books such as *Human Diversity*[6] by the conservative provocateur Charles Murray, eugenic thinkers have implied a specific set of answers to these questions: First, that the causal chain between genetics and social inequality is short and primarily mediated via the development of intelligence. Second, that the causal chain between genetics and social inequality is *best* understood at a cellular and organismic level of analysis, with intelligence seen as an inherent property of a person's brain, rather than as something that develops in a social context. And, third, that the alternative possible worlds where this chain is broken are dystopian, requiring either massive state intrusion into people's home lives or widespread genetic engineering. In short, the eugenic formulation is that genes cause social class the way genes cause Huntington's, via mechanisms that are universal, intuitively biological, and difficult (if not impossible) to modify.

Conceptualizing the links between genetics and social inequality in terms of a short, biological, and universal causal chain saps political will to address inequality. As the philosopher Kate Manne put it, "naturalizing" social inequalities serves the function of "making them seem inevitable, or portraying people trying to resist them as fighting a losing battle."[7] To say that government should do something to redress inequality is to imply that change is possible, whereas a strict genetic determinism suggests that change is impossible—so why bother? And the idea that the relationship between genes and social inequality is best understood at the level of a person's cellular biology, rather than at the level of how societies organize themselves, resonates with the eugenic notion that some people are just inherently *better* than other people. As the evolutionary biologist Theodosius Dobzhansky summarized back in the 1960s, "the favorite argument of conservatives has always been that social and economic status merely reflects intrinsic ability."[8]

Such ideologically motivated talking points about the mechanisms linking genes and social inequalities can obscure the science itself. Ultimately, neither eugenic ideas about human superiority nor thought experiments about red-headed children is a substitute for empirical results: What *do* we know about the mechanisms linking genes with social inequalities, particularly inequalities in educational outcomes?[9]

This field is moving fast, but for now, I think we can say five things about the mechanisms linking genetics to inequalities in education:

1. The genes relevant for education are active in the brain, not in the hair or skin or liver or spleen.
2. The mechanisms linking genes to education start very early in development, before a child is even born.
3. Genetic effects on educational success involve the development of the types of intelligence that are measured by standardized tests . . .
4. . . . but not just intelligence. Genetic effects on educational success also involve the development of what are called "non-cognitive" skills.
5. Understanding the mechanisms of genetic effects requires understanding the interactions between people and their social institutions.

Now, let's consider each of these five points in detail.

The Question of Where: Genes Affect the Brain

As I described in chapter 3, a GWAS of educational attainment or any other phenotype gives you a relatively minimal set of results, with a list of SNPs and how strongly associated they are with the outcome you are studying. On its own, this set of results doesn't tell you much of anything about mechanism. But, like a Talmudic author annotating the slim text of the Torah, *bioannotation* analysis takes the minimal results produced by GWAS and provides explanatory notes based on what is known about genomic and cellular biology: Individual SNPs are mapped to genes; genes are mapped to gene functions and products, like proteins; gene products are mapped to biological systems within cells or tissues.

One important tool in the bioannotation toolbox is to test whether the genes that are associated with an outcome are preferentially *expressed* in certain parts of the body or in certain cell types. Every cell in your body has the same DNA code, but different cells need to do different things, so they turn different genes on and off in characteristic patterns of gene expression. Given a set of GWAS results, then, analysts can test whether the genes that are most associated with a trait like educational attainment or subjective well-being or obesity are also more likely to be expressed in one part of the body than another.

This type of gene expression work has yielded an important insight: genes associated with educational attainment are preferentially expressed in the brain, and within the brain, they are preferentially expressed in neurons. Zeroing in on the "top" genes associated with educational attainment, they are involved in the processes that are critical to the ability of neurons to communicate with one another. Those processes include the secretion of neurotransmitters that carry messages from neuron to neuron, the plasticity of neuronal connections in response to new information or in response to disuse, and the maintenance of ion channels that are necessary for a neuron's electrical charge. The centrality of the brain in terms of gene expression is also seen for every other phenotype relevant to social inequality—subjective well-being and depression, alcohol use and smoking, obesity and income.

Returning again to Jencks's example of red-headed children, the heritability of educational attainment could have been picking up on genetically caused differences in physical appearance, which then elicited differential treatment from others. If that were the case, however, then the genes associated with educational attainment would be expressed in places in the body *other* than the brain. But that's not what we see. Whatever genes are doing to make it more or less likely for some people to succeed in education, they are doing it in people's brains, not their hair or livers or skin or bones.

The Question of When: Genes Start Their Effects Very Early in Development

Another piece of information that bioannotation analyses can pull in is *when* in development the genes that are associated with an outcome are expressed. Different genes are active in our bodies at different points in our life span. Genes relevant for growth, for instance, are necessary when the body is rapidly growing in size, but not so relevant once adult stature has been obtained. This type of analysis has revealed that some of the genes associated with educational attainment are preferentially expressed prenatally, while a child's brain and nervous system are still being formed.[10]

A different strategy for understanding *when* genes become relevant is to analyze data from twins who are measured at different ages. My colleagues and I, for instance, looked at a sample of twins who had been measured on their cognitive abilities very early in life—at age 10 months and age 2 years.[11] Tests of cognitive ability at such early ages ask children to do things like repeat sounds, put three cubes in a cup, or pull a string to ring a bell. At 10 months, there were no apparent genetic effects on differences in measured cognitive abilities, but genetic effects emerged by the time children were 2 years old.

Other research has used polygenic indices created from the GWAS of educational attainment in order to see *what* phenotypes are correlated with polygenic indices, and *when* in development these correlations are apparent. This work has shown that the education polygenic index is correlated with whether children start talking before age 3 and their scores on IQ tests at age 5.[12] So, consistent with what was observed in bioannotation and twin studies, polygenic index analyses suggest that, whatever genes are doing to influence educational inequalities, they are doing it early in life—with effects that are apparent before children ever begin school.

The Question of What: Genetic Effects Involve Basic Cognitive Abilities

In the twin study that I co-direct at the University of Texas, we measure a set of cognitive abilities known as *executive functions*. Over the course of several hours, children complete twelve different tests (illustrated in figure 7.2).

Although the term "executive functions" is plural, children who tend to do well on a test of one executive function tend to do well on all the other tests. This positive correlation among test scores

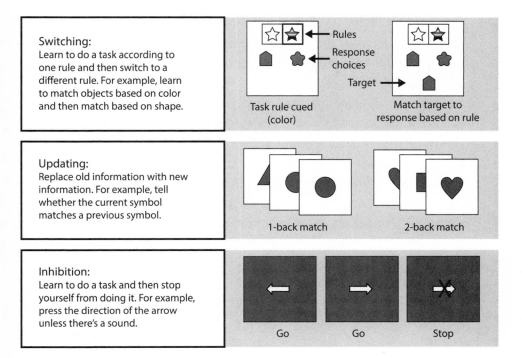

FIGURE 7.2. Examples of tests of executive functions in children. Described in Laura E. Engel-hardt et al., "Genes Unite Executive Functions in Childhood," *Psychological Science* 26, no. 8 (August 1, 2015): 1151–63, https://doi.org/10.1177/0956797615577209.

means that performance on all of them can be aggregated statistically into a single overall score that we call *general EF*. Children who have higher general EF are better at regulating their attention. They can stop themselves. They can shift from one rule to another. They update information in real time and keep small amounts of information accessible in their working memory.

Two things about general EF fascinate me. First, it is nearly 100 percent heritable.[13] That is, within a group of children who are all in school, nearly *all* of the differences in general EF between them are estimated to be due to the genetic differences between them. We have tested EF abilities in hundreds of 8- to 15-year-old twins, and after you correct for measurement error (the tendency for scores on tests to vary slightly because of randomness), identical twins are essentially *exactly the same* in their general EF abilities. Fraternal twins are correlated at 0.5—they share half as much genetic variance;

they are half as similar to one another. Nearly perfect heritability is rare for any behavioral trait, particularly one measured in childhood. General EF is as heritable as eye color or height, more heritable than BMI or pubertal timing.[14]

Second, this nearly perfectly heritable trait is a surprisingly good predictor of how well students do on their state-mandated academic achievement tests. Like public schoolchildren all around America, students in Texas are required to take standardized tests of mathematics and reading skills at the end of the school year, beginning in grade 3. We received school transcripts for the children who participated in our study, so we could see if their performance on our in-lab EF tests predicted their performance on the high-stakes tests given by their schools. It did: general EF was correlated at 0.4 to 0.5 with students' test scores.

This is just one study that illustrates a more general pattern. Twin studies have long found evidence for genetic influences on basic cognitive abilities.[15] These abilities, which are typically measured in highly controlled laboratory testing situations, in turn predict better performance on all sorts of tests, like state-mandated achievement tests in primary school and admissions tests that gate entry to college and graduate school.

This pattern is also evident in studies that use polygenic indices rather than twins to study genetic influence. As I described in the previous section, if you conduct a GWAS of educational attainment in adulthood, and then use the results of that GWAS to create a polygenic index, that polygenic index is also associated with children's performance on an IQ test as early as age 5. Education polygenic indices are also associated with how well children read at age 10, their IQ scores at age 13, and their scores on university admissions tests at age 17.

At every point in formal education, people who can memorize facts quickly, easily redirect their attention, and manipulate abstract information in their head *do better on tests*. And, whether a person performs well on tests is a major factor that gets you advanced to the next stage of your schooling.

The Question of What, Revisited:
Genetic Effects Involve More than Intelligence

As Dostoevsky reminded us, "It takes something more than intelligence to act intelligently."[16] The journalist Paul Tough, in his best-selling book, *How Children Succeed: Grit, Curiosity, and the Hidden Power of Character*, argued that "some children succeed while others fail" because the former have "character" traits such as "perseverance, curiosity, conscientiousness, optimism, and self-control."[17] The Nobel laureate James Heckman gave a similar list: "motivation, perseverance, and tenacity are also important for success in life."[18] This constellation of traits is often labeled social-emotional skills, or more generally as "non-cognitive" skills.

The label "non-cognitive" is a misnomer: behavioral control and interpersonal skills are obviously brain-based, cognitively demanding phenotypes. But the "non" in "non-cognitive" serves to emphasize what these motivational, behavioral, and emotional traits are *not*—they are not synonymous with performance on standardized tests of cognitive ability or academic achievement.

Psychological research on non-cognitive skills was popularized by books like *How Children Succeed* and Angela Duckworth's *Grit: The Power of Passion and Perseverance* (both *New York Times* best-sellers), and by TED talks such as Dr. Carol Dweck's on mindset (viewed more than 12 million times).[19] As words like "grit" and "growth mindset" entered the popular lexicon, conjecture about the role of genetics in their development quickly outpaced science, with many commentators quick to position such skills in opposition to genetics. For instance, Tough wrote, "The character strengths that matter so much to young people's success" are not "a result of good luck or good genes."[20] Similarly, Jonah Lehrer (whose work has now been discredited for plagiarism and fabrication) wrote an article for *Wired* magazine on "the importance of grit" that portrayed grit as a counterweight to the importance of genetic influence: "The intrinsic nature of talent is overrated—our genes don't confer specific gifts. . . . Talent is really about deliberate practice."[21]

FIGURE 7.3. Different types of non-cognitive skills. Described in Elliot M. Tucker-Drob et al., "Genetically Mediated Associations between Measures of Childhood Character and Academic Achievement," *Journal of Personality and Social Psychology* 111, no. 5 (2016): 790–815, https://doi.org/10.1037/pspp0000098.

I speculate that part of the public enthusiasm for non-cognitive skills is that they were presumed to be free of the genetic influences that bedevil conversations about cognitive ability. But this presumption is incorrect. There are three lines of evidence that suggest that the development of non-cognitive skills is *part* of the pathway connecting genes to educational outcomes.

First, we can study non-cognitive skills in twins. In the twin study that I co-direct in Texas, we designed a battery of measures to try to capture a breadth of traits thought to be important for success in school and beyond (figure 7.3). These include the "greatest hits" of the past few decades of social and educational psychology, including grit, growth mindset, intellectual curiosity, mastery orientation, self-concept, and test motivation. In our twin sample, non-cognitive skills are moderately heritable (around 60%), an estimate that is consistent with what most groups have found for IQ (50% to 80%).

Second, researchers have taken polygenic indices created from GWAS of educational attainment and seen what phenotypes, *other than* cognitive test performance, those polygenic indices are correlated with in childhood and adolescence. This work has found that "education" polygenic indices are correlated with:[22]

- How interpersonally skilled ("friendly, confident, cooperative, or communicative") children seemed to adults at age 9
- How often children were truant from school at age 11
- How likely teachers were to say that children had ADHD symptoms at age 12

- How much teenagers aspired, at age 15, to someday work in a high-status profession, like medicine or engineering
- How agreeable and open to new experiences people were in childhood and adulthood.

A third approach to studying the genetics of non-cognitive skill is to study people who differed in their educational attainment but who were similar in their performance on tests of cognitive ability. This approach allows you to ask: After you take out cognitive ability from educational attainment, what is left over? My colleagues and I borrowed this strategy and adapted it to GWAS, testing which SNPs are associated with differences in how far people go in school (higher educational attainment), *above and beyond* their association with cognitive test performance.[23] The result is a set of GWAS results for "non-cognitive" variation in educational attainment.

An early follow-up study of these GWAS results compared siblings who differed in their "non-cognitive skills" polygenic index and found evidence that our GWAS was indeed tapping genes that are causally related to educational success.[24] Additionally, we used these GWAS results to calculate what are called "genetic correlations" with a variety of other traits. Genetic correlation analysis uses results from GWAS of two different traits to estimate the strength of the relationship between the genes influencing each trait.[25]

We found that the genetics of non-cognitive skills related to greater educational attainment were associated with a wide variety of different types of things.[26] In the domain of personality, non-cognitive genetics were most strongly related to a trait called Openness to Experience, which captures being curious, eager to learn, and open to novel experiences. The genetics of non-cognitive skills were also correlated with the ability to defer gratification, as measured by people's preferences for larger, later rewards over smaller, immediate rewards; with later childbearing; and with less risk-taking behavior generally. Overall, our results suggest that non-cognitive skills really are skills, plural—many different genetically associated traits and behaviors contribute to going further in school.

There were also some surprises. The SNPs correlated with non-cognitive skills were correlated with *higher* risk for several mental disorders, including schizophrenia, bipolar disorder, anorexia nervosa, and obsessive-compulsive disorder. This result warns us against viewing the genetic variants that are associated with going further in current systems of formal education as being inherently "good" things. A single genetic variant might make it a tiny bit more likely that someone will go further in school, but that same variant might also elevate their risk of developing schizophrenia or another serious mental disorder.

Overall, these three lines of research show us that non-cognitive skills are *not* a get-out-jail-free card for grappling with the implications of genetic influence on inequality-related traits. Rather, part of the reason *why* one's genotype comes to be correlated with one's ultimate educational attainment is that motivation, curiosity, interpersonal skills, and persistence are themselves genetically influenced, and these traits promote greater success in school.

The Question of Who: Genetic Effects Involve Interactions Among People

As I described in point #2, genetic effects are apparent very early in life: genes identified in a GWAS of educational attainment are expressed as early as the prenatal period and are associated with performance on IQ tests as early as age 5. At the same time, we also observe a pattern that might seem unintuitive: genetic effects on cognitive abilities, in particular, only get stronger over time. One meta-analysis (a type of study that pulls together and summarizes data from lots of different individual studies) found that genetic effects on cognitive ability rapidly get stronger from birth until the end of childhood, around age 10.[27] A similar increase in the strength of genetic effects on personality traits like orderliness and openness to new experiences is also evident, but over a longer period of time, with genetic effects increasing until people are around age 30.

Why might genetic influences get stronger over time, even as children are accruing more and more environmental experiences? The secret to understanding this apparent paradox is to understand that *interactions with the social environment are an essential part of the causal chain connecting genetics to psychological and social outcomes.*[28] Intelligence, curiosity, motivation, self-discipline: these do not emerge in a vacuum as some "inherent" or "inborn" property of a person's nervous system. Rather, they unfold over time, as part of a reciprocal dance between children and the people in their lives.

In one of our first studies on this idea, we examined a sample of twin children who were 4 years old and their parents.[29] Parenting behaviors were measured by filming parents (usually mothers) interacting with their kids for ten minutes with two bags of toys. Trained raters were instructed to rate parents on their levels of *cognitive stimulation*—did the parent try to teach the child things that would enhance their verbal or perceptual development, and did the parent do that in a way that was developmentally appropriate and tracked the child's interests?

We found two main results. First, children who had more-advanced cognitive function at age 2 received *more* cognitive stimulation from their parents at age 4, even controlling for the parent's previous parenting behavior. Second, parents who provided more cognitive stimulation to their children at age 2 had children who had better reading skills at age 4, even controlling for their previous levels of cognitive function. This study is giving us insight into an early part of the causal chain: children who, at age 2, are better able to repeat sounds and sort toys are responded to *differently* by their parents than children who don't babble back. And the cognitive stimulation provided by parents makes a difference to how well children read at age 4. An initial genetic advantage in the ability to repeat sounds early in life, then, is passed down the causal chain via the type of parenting behavior a child receives.

Another study using polygenic indices rather than twins examined several different aspects of the home environment when children were young and tested which aspects were associated with the *parents'* genetics, as measured by their polygenic index for education.[30] Interviewers went into the home and measured how warm and affectionate parents were, how safe and tidy the house was, how chaotic and disorganized the house was, and how cognitively stimulating the parents were. Of these, only cognitive stimulation—measured by the availability of toys, puzzles, and books, and by activities done with the parent, like going to the zoo or museum—was associated with both the parents' genetics and how far children ultimately went on in school.

The importance of interactions with the social environment doesn't stop in early childhood. My colleagues and I did a study using a sample of around 3,000 Americans who were all high school students in 1994–1995.[31] This sample is interesting because they both gave their DNA to be genotyped and released their high school transcripts so that researchers could see what courses they had taken every year. Combining these different sources of information, we could see how the process of curricular tracking in American high schools might be part of the causal chain connecting genetic differences between students and ultimate educational outcomes.

When they start high school, most students have a choice about what math class they are going to take. Depending on a number of factors—what math class they took in the eighth grade, how much they like and are interested in math, what their teacher and school counselors think of them, how much their parents are knowledgeable about what classes are necessary to get into college—students are either placed in algebra (the most common math class for ninth graders in America), or in a remedial class like pre-algebra, or they take an "advanced" class like geometry.

When I give a talk to other academics about math achievement in high school, I ask the audience to raise their hand if they took calculus in high school. Almost without exception, every hand in the room goes up. Most people who eventually get a PhD in a STEM discipline had a fairly advanced level of mathematics training in high school.

But calculus training in high school is actually pretty rare. In 2018, about 15 percent of high school seniors completed calculus, and even fewer students completed calculus back in the 1990s, when most states required only two years of math to graduate high school.[32] In the sample we used for our study, 44 percent of students who were enrolled in geometry in the ninth grade eventually took calculus, compared to just *4 percent* of students enrolled in algebra in the

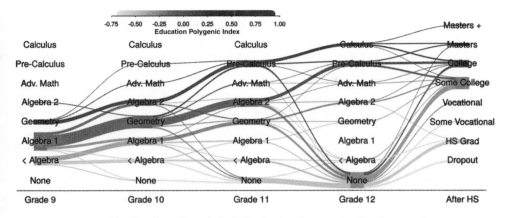

FIGURE 7.4. Flow of students through the high school math curriculum by educational attainment polygenic index. Width of the line represents number of students enrolled in each math course in each year of high school (secondary school). Darkness of the line represents the average education polygenic index of students enrolled in that course. Values of the polygenic index are in standard deviation units. Data are from European-ancestry students from the National Longitudinal Study of Adolescent Health who were enrolled in US high schools in the mid-1990s. Reproduced from K. Paige Harden et al., "Genetic Associations with Mathematics Tracking and Persistence in Secondary School," *Npj Science of Learning* 5 (February 5, 2020): 1–8, https://doi.org/10.1038/s41539-020-0060-2.

ninth grade. When fourteen-year-olds are making a decision about what math class to take, they probably don't know that, if they pick the harder class, they will have *18 times greater odds* of having the opportunity to learn math skills that are critical for a future in STEM.

In the United States, we don't prohibit redheads from going to school, but we *do,* in practice, prohibit students who didn't take geometry in the ninth grade from having the opportunity to learn calculus in high school, and we prohibit students who haven't completed Algebra 2 from graduating from high school (in many US states) or from enrolling in flagship public universities.

Figure 7.4 visualizes the flow of students through the high school math curriculum as a function of their genes. From top to bottom are different math courses ordered by level of difficulty, ranging from basic/remedial math to calculus. From left to right are the four years of high school (secondary school) and the student's ultimate educational outcome. The width of the rivers from year to year represent the number of students following that particular curricular trajectory. And the darkness of the river represents the students' polygenic indices derived from the GWAS of educational attainment.

Here, we see that students are separated into different opportunities to learn, based on characteristics that are themselves genetically influenced, leading to genetic stratification in education at the beginning of high school—the polygenic river is already darker for students in geometry than for those in pre-algebra. From there, we see a remarkable path dependence, with subsequent course-taking depending on previous course-taking. At the same time, students with low polygenic indices are more likely to drop out of math at every year, leading to an increase in genetic stratification as high school goes on.

Again, we should revisit the word "mechanism." Colleges and universities cannot see a student's DNA when he or she applies to college. They can see, however, the student's transcript, which gives information about whether he or she has completed the "college-ready" courses, including Algebra 2, necessary for admission. If curricular tracking processes result in students' genotypes becoming correlated with accumulating (or not accumulating) certain math credentials that university admissions offices are paying attention to, then the curricular tracking process and the university admissions process become mechanisms for genetic effects. The ways institutions assign students, promote students, and admit students transmute invisible DNA into visible academic credentials.

Red-Headed Children, Redux

When Jencks first proposed his thought experiment of red-headed children in the 1970s, it was becoming clear that genes did have an effect on academic achievement, intelligence, income, psychopathology, health, and well-being, but—as he correctly pointed out—those genetic effects could be transmitted via any number of mechanisms. In the fifty years since, however, we have—thankfully—learned some things. Thousands upon thousands of genetic variants matter for educational attainment and other complicated human phenotypes. These genes exert their effects via largely unknown cellular processes that are happening in neurons and other brain cells. These cellular effects are already happening during prenatal development, and their effects on the individual organism are already evident in childhood in terms of better early vocabulary, higher executive functions, and stronger non-cognitive skills. Children who have larger initial advantages in these skills are responded to differently by parents and by educators: they are given more cognitive stimulation at home and they are given more challenging coursework at school, both of which compound their initial advantages. And this entire process plays out over years, in the context of a test-heavy formal educational system.

So, what about the other question implied by the red-headed child example: In which alternative possible worlds would these causal chains be broken, and are these alternative possible worlds that we would like to live in? Considering this question leads us to the second half of the book.

By this point in the book, I hope I have convinced you of three things. One, genetic research has developed an array of methods, using family members, measured DNA, and combinations of both, that estimate the effects of genes on complicated human outcomes. Second, the overwhelming consensus of that research is that genetic differences between people matter for who succeeds in formal education, which structures many other forms of inequality. Third, while the biology of these genetics is still largely a mystery, progress is being made on understanding the psychological and social mediators of genetic effects on educational success. Now, let us turn our attention to how these insights should be used—in policy, in educational practice, in re-examining our myths about meritocracy.

PART II

Taking Equality Seriously

Alternative Possible Worlds

Genetic studies ultimately hope to answer a question about one set of alternative possible worlds: Given that you were born and raised in a particular time and a particular place, what if you had inherited different genes? In contrast, Sandy Jencks's thought experiment about red-headed children, which I described in the last chapter, is asking a *different* question, about a different set of alternative possible worlds: What if your genotype were the same, but the social and historical context changed?

That is not just a rhetorical question. In 1989, when the Berlin wall fell, Philipp Koellinger, the lemon-chicken-loving economist that you met in chapter 3, was fourteen years old. He had, up to that point, spent his entire life in East Berlin, but after the wall came down, Koellinger and other East German students had access to a whole new world of educational opportunities. Governments fall, borders dissolve, economies change, laws are passed, and policymakers change their minds. Societies are reimagined and remade.

Since Francis Galton, eugenic thinkers have steadily and successfully engaged in a misinformation campaign, convincing people that the reimagination of society is futile. Their propaganda is this: if genetic differences between people cause differences in their life outcomes, then social change will be possible only by editing people's genes, not by changing the social world. This was the thesis of a bombshell paper written in the late 1960s by the psychologist Art Jensen: "How much can we boost IQ and scholastic achievement?," he asked, and he used early research on the heritability of academic achievement to answer very much in the negative.[1]

Fast-forward several decades, and the writer Charles Murray has continued the same drumbeat of hereditarian pessimism. He argued in *Human Diversity* that "outside interventions are inherently constrained in the effects they can have on personality, abilities, and social behavior"[2]—inherently constrained *because* these aspects of ourselves are genetically influenced. In this view, people are seen as having an inherent genetic "set point," with a small amount of environmentally induced jitter around that set point. Social change, Murray thinks, can potentially affect that small amount of variation around the genetic set point, but cannot budge the set point itself.

This hereditarian pessimism about the possibility of social change, however, is based on a fundamental misunderstanding of the relationship between genetic causes and environmental interventions. As the economist Art Goldberger quipped in the late 1970s, your genetics caused your poor eyesight, but your eyeglasses still work just fine.[3] That is, eyeglasses don't just help with the environmentally caused portion of bad eyesight. They help with all of your eyesight, regardless of whether it is genetically or environmentally caused. In so doing, they serve as an outside intervention that severs the association between one's myopia genes and having functional vision.

The eyeglasses example is instructive about a more general point. The answer to one "What if?" question—What if, all other things being equal, you had inherited a different combination of genes from your parents?—does not imply anything straightforward about the answers to another "What if?" question—what if, your genotype being exactly equal, the social and economic world were changed?[4] Would Koellinger's probability of getting a PhD been different if he had inherited a different combination of genetic variants? Yes. We know this is true from sibling comparisons of polygenic indices and from twin and measured DNA studies of heritability. But even if there are genetic causes of educational outcomes, would Koellinger's probability of getting a PhD have been different if the Berlin wall had not come down? Also yes. Heritable phenotypes are not immune from social change.

Unfortunately, the mistaken idea that genetic influences are an impermeable barrier to social change is also widely endorsed not just by those who are trying to naturalize inequality, but also by their ideological and political opponents. Theodosius Dobzhansky, the Russian-born evolutionary biologist who in the wake of World War II sounded the alarm about Stalin's persecution of geneticists, remarked on this irony in 1962: "Oddly enough, some liberals come close to agreeing with diehard conservatives, that if it were shown that people are genetically diverse then attempts to ameliorate their lot by social, economic, and educational improvements would be futile, and perhaps even 'contrary to nature.'"[5]

Dobzhanky's characterization of the response to genetics remains remarkably prescient. Consider, for example, the anthropologist Agustin Fuentes, who encapsulated this attitude in an interview he gave for the documentary film, *A Dangerous Idea:*[6] "If you believe that someone's ability to do well as a captain of industry . . ., if you believe that that's written in the DNA in some way or another, then you have no responsibility, and things can stay the way they are." Fuentes implies that people have a moral responsibility to work for a more egalitarian society. Accordingly, he rejects the idea that social inequality is "written in the DNA in some way or another," lest that idea interfere with people's advocacy for and investment in social change.

But both things can be true at the same time: genetics can be causes of stratification in society, *and* measures to address systematic social forces can be effective at enacting social change. Once you have a clear understanding of this dual truth, a huge part of the controversy surrounding behavioral genetics dissipates, leaving space to address two much more interesting—and much more complicated—questions. First, how *have* social and historical contexts differed in ways that, like putting on eyeglasses, change the relationship between genotype and phenotype? Second, looking forward to the question of policy, what do we *want* the relationship between people's genetics and their outcomes to look like? These are the questions that we'll consider in this chapter.

Leveling Down: When the Worst Environments Produce the Most Equal Outcomes

Koellinger and the other children of East Germany were of course not the only ones who saw their educational opportunities change with the collapse of the Soviet Union. Estonia is a Baltic state that was occupied by the Soviets from the end of World War II until 1991. During the Soviet occupation, students had few free choices.[7] At the end of the eighth grade, they were assigned to one of three school tracks, with minimal movement among them. Upon completing their education track, students were then assigned to a workplace where they had to work for at least three years. A university degree was not particularly prized, and there was little competition for admission to university.

The end of the Soviet era allowed for free choice and competition in education and in jobs. Now, Estonians enjoy what the Organisation for Economic Cooperation and Development (OECD) has called a "high performing education system" that "combines equity with quality." Like a handful of other OECD countries (e.g., Finland, Norway, Korea, and Iceland), Estonia has above-average scores on tests of reading, and little of the variation between students within the country is explained by student socioeconomic background.[8]

In addition to developing a high-quality, high-equity educational system, Estonia also boasts one of the best national biobanks in the world. The Estonian Genome Center has been amassing a large-scale database about the Estonian population, including information about their health and their genes. Some of the people in the Estonian biobank came of age under the Soviets. Some of them came of age after the fall of communism. So, in 2018, geneticists from the UK posed the question: What happens to genetic causes when society changes?

Specifically, they created a polygenic index from a GWAS of educational attainment and tested its relationship with educational attainment in people who were younger than 10 years old when the Soviet occupation ended (i.e., before assignment to secondary school tracks ended) versus the rest of the sample. What they found is that the polygenic index accounted for significantly more variance in the post-Soviet group, compared to those educated in the Soviet era. When children are assigned to schools with little regard to choice or competition, the genetic differences among them are more weakly related to their ultimate educational outcomes.

Similar results are evident when examining polygenic index associations vis-à-vis historical changes that have allowed women greater educational opportunity in the United States.[9] For my grandmother's birth cohort (people who were born in 1939–1940), the polygenic index was more weakly related to educational attainment among women than among men. (These women were in their thirties before my alma mater, the University of Virginia, admitted students without regard to gender, in 1972.) But this gender difference has narrowed over time: as educational opportunities for women increased, the polygenic index has become more strongly associated with women's educational outcomes. For woman in my birth cohort (people born in 1975-1982), the polygenic index is as strongly associated with education as it is for men. Genetics, ironically, has become a sign of gender equality.

Evidence of the same pattern—higher genetic associations within social contexts that allow for greater choice and competition— emerges when analyzing twin data. An early (1985) twin study in Norway[10] found that the heritability of educational attainment was higher for later-born cohorts, particularly the men, who benefitted from educational reforms expanding access to higher education, than it was for earlier-born cohorts.

In addition to comparing twin heritability within a country across time, one can also compare heritabilities *across* countries that differ in their intergenerational social mobility, defined by the parent-child correlation in years of schooling.[11] Despite the mythology of the United States as the "land of opportunity," it has lower social mobility than many other countries; Denmark is an example of a country with high social mobility. The heritability of educational attainment is actually *lower* in countries with lower social mobility, like the United States and Italy. This study reminds us that heritability is about differences—and that even family members differ in their genetics. In a more static society, where education is reproduced from parent to child with little movement upwards or downwards, the genetic lottery makes less of a difference to a child's life outcomes. In contrast, when life opportunities depend less on a family's level of financial and cultural resources, genes can make more of a difference.

Finally, twin studies have shown that the heritability of child cognitive ability is *lowest* for children raised in poverty and highest for children from rich homes—particularly in the US, where social safety nets for poor families are weaker than in other countries.[12] The causal chain from genes to performing better on an intelligence test is not entirely broken, but it is weakened, when children have few material resources in their homes.

Together, these studies illustrate a process of *leveling down*: people were prevented, by poverty or by sexism or by repressive government, from continuing their educations, rendering the genes they have largely irrelevant. The causal chain between genotype and going to school depends on having a school to go to. Thus, we often see that the social contexts in which genetic effects on education are minimized are the *least* desirable ones, as they involve deprivation, discrimination, and/or authoritarian social control.

This pattern of results—where heritability is higher in good environments than in bad ones—can be counterintuitive. But we can build up an intuition about them using a classic thought experiment from the biologist Richard Lewontin.[13] Imagine two gardens. One has nutrient-rich soil, bright sunlight, and plenty of water, whereas the other is rocky, dark, and parched. Now imagine that both gardens are sown with genetically diverse corn seeds. Within the lavishly resourced garden, each plant has the opportunity to reach its maximal height. Moreover, because the conditions of the garden are exactly uniform across plants, the variation among the plants in their heights will be primarily due to genetic differences between the seeds.

Lewontin's example is often invoked to illustrate why, as I explained back in chapter 4, differences between groups (such as racial groups) might be entirely caused by environmental factors, even when differences within groups are caused by genetic differences.[14] At the same time, the garden example also illustrates a point that is often lost in the rhetoric about "closing gaps" between students in education:[15] the well-resourced garden, which provides identical environments to all its plants, might have taller plants on average, but also plants that are more unequal in their heights. Similarly, removing structural barriers, such as institutionalized gender discrimination, unaffordable tuition costs, and strict tracking systems, can both increase the *average* level of education in a population and also increase the amount of *inequality* in educational outcomes that are associated with genetic differences between people.

Equality versus Equity

As we've seen, empirical studies often observe lower heritability of educational outcomes in repressive and deprived environments, and higher heritability in more open and well-resourced ones. On the basis of this observation, some scientists have proposed that high heritability is actually a *good* thing—a sign that grossly damaging environmental conditions have been ameliorated, and a sign that society is treating individuals *qua* individuals, such that each person's unique genetically influenced talents and predispositions shine forth and affect their life outcomes. In the 1970s, Richard Herrnstein, in his book *IQ in the Meritocracy*, noted that high heritability was a positive sign that society had eliminated some environmental inequalities: "Eliminating large classes in school, poor libraries, shabby physical surroundings, teeming ghettos, undertrained teachers, inadequate diet, and so on . . . have the corollary effect of increasing heritability."[16] More recently, the social scientists Dalton Conley and Jason Fletcher returned to this point, suggesting that high heritability might be considered a "measure of fairness" in society, a "necessary—but not sufficient—component of a utopian society of equal opportunity."[17]

This suggestion, that high heritability of life outcome is a necessary component of a utopian society, will strike some readers as profoundly unsettling. Is this truly our ideal—a "genetic Shangri-la"[18] where inequalities of life outcome due to poverty and oppression have been removed, but inequalities of life outcomes due to genetics remain?

Why are inequalities that are related to your genes more acceptable than inequalities rooted in the social circumstances of your birth? After all, as I've argued throughout this book, both are accidents of birth, forms of luck over which a person has no control.

In the early 1980s, Leon Kamin, a psychologist who was a fierce critic of behavioral genetics, pushed back against the intuition that genetically caused inequalities of life outcome were any more morally acceptable than environmentally caused ones. His example was phenylketonuria (PKU), a rare disorder caused by a single-gene mutation that impedes the body's ability to metabolize a protein building block called phenylalanine. Untreated, PKU causes intellectual disability. But high-income countries now routinely screen newborns for PKU, which is treated with a restricted diet low in phenylalanine.

The treatment of PKU with diet, just like the treatment of myopia with eyeglasses, reminds us of a point I was making earlier in this chapter: *genetic causes can have environmental solutions*. In fact, despite the fact that PKU has a simple and well-understood genetic etiology, environmental solutions currently remain the *only* solutions. Gene therapy for PKU is not (yet) a reality.[19] And we can contrast the simple etiology of PKU with the genetic architecture of highly polygenic outcomes, like intelligence test scores or educational attainment, which involve thousands upon thousands of genetic variants with tiny effects and unknown mechanisms.[20] To make matters even more complicated, many of these variants are also involved in phenotypes that are valued differently by society: many of the same genetic variants associated with higher educational attainment, for instance, are also associated with higher risk for schizophrenia.[21] The suggestion from some conservative academics that we might edit children's genomes to increase their IQs is not just scientifically unfeasible; it is scientifically absurd.[22]

And, as Kamin pointed out, the example of PKU also shows up the absurdity of being concerned only about environmentally rooted inequalities, but not genetically rooted ones:[23]

> Why should a liberal *not* be upset if there are long-term *genetic* affects [*sic*] of families on their offsprings' life chances? Are "genetically" produced differences more just, good, or true than "environmentally" produced differences? . . . Are "genetic" differences more fixed and irreversible than "environmental" differences? To argue that we should be upset by cultural-familial retardation, while cheerfully accepting the genetically determined (but easily preventable) PKU, would be obviously absurd.

Kamin's rhetorical questions—"Why should a liberal *not* be upset?"—mirror the arguments of the political philosopher John Rawls, who pointed out that if one found inequalities stemming from environmental luck disturbingly unfair, one might also find inequalities stemming from genetic luck just as disturbing:[24]

> Once we are troubled by the influence of either social contingencies or natural chance on the determination of distributive shares, we are bound, on reflection, to be bothered by the influence of the other. From a moral standpoint the two seem equally arbitrary.

The idea that we should be bothered by inequalities that are due to factors over which people have no control, and that they therefore cannot be said to deserve, has not stayed in the world of abstract philosophy. Rather, it is now baked into education policy around the world. Consider how the OECD defines *equity* in education:[25]

> Equity does not mean that all students obtain equal education outcomes, but rather that *differences in students' outcomes are unrelated to their background or to economic or social circumstances over which students have no control* (emphasis added).

The logic here is plainly stated. Any inequalities between students who come from different social classes are considered unfair, and the reason *why* they are unfair is that they are due to the luck of the draw, rather than under the choice or control of the student.

FIGURE 8.1. Equality versus equity. Image from Interaction Institute for Social Change. Artist: Angus Maguire.

This vision of what equity means has deeply penetrated the thinking of educators in America.[26] A memetic illustration of the difference between equity and equality features three people of different heights, all trying to peer over a baseball fence (figure 8.1). "Equality" is them each getting a stool of the exact same height, resulting in the persistent expression of individual differences among them. In contrast, "equity" is illustrated as each person getting a stool high enough for them to see over the fence, with higher stools (i.e., more intensive support) given to the shortest people.

Instead of treating everyone the same, then, equity in education is thought to involve giving the children who are most likely to struggle in school (either because of their background social conditions or because of "natural chance") tailored and intensive supports to bring their learning, as much as possible, up to the level more easily attained by their more advantaged peers. Reinforcing the idea of equity in terms that a five-year-old can understand (figure 8.2), my own daughter's pre-K classroom featured a bubble-lettered, rainbow-colored sign stating that "Fair isn't everybody getting the same thing. Fair is everybody getting what they need in order to be successful."

Advocates of an equity perspective object to the rhetoric about "equality of opportunity" that dominates American political

FIGURE 8.2. Pre-kindergarten classroom sign about fairness. Photo by author.

discourse. Equality of opportunity can actually be defined in multiple different ways,[27] but the most straightforward definition is simply to treat everyone exactly the same. The problem, of course, is that people are not exactly the same, genetically and otherwise. Like building a fence that is six feet tall and giving everyone, regardless of their height, the exact same six-inch footstool to see over it, an education system that provides everyone with equality of opportunity in the form of exact uniformity of educational conditions will ineluctably produce profound inequalities of outcome.

The fact that equal opportunity will necessarily reproduce inequalities that are rooted in the arbitrariness of nature has led some to encourage abandoning our attachment to the idea. The

philosopher Thomas Nagel observed, "The liberal idea of equal treatment . . . guarantees that the social order will reflect and probably
magnify the initial distinctions produced by nature and the past . . .
[T]he familiar principle of equal treatment, with its meritocratic
conception of relevant differences, seems too weak to combat the
inequalities dispensed by nature."[28] The writer Freddie deBoer put
it more trenchantly: "Equality of opportunity is a shibboleth. It's a
ruse, a dodge. It's a way for progressive people to give their blessing
to inequality."[29]

Raising the Floor: When Interventions Promote Equity

To return to Goldberger's example of eyeglasses, we can see that
part of why this thought experiment has been so enduring is that
eyeglasses are an equity-promoting intervention. People with good
eyesight are not being given vision-enhancing surgery to make their
vision extra acute. Rather, resources are being selectively given to
those with poor eyesight to bring their daily functioning, as much
as possible, up to the level enjoyed by those with good eyesight.

More cynically, however, I suspect that Goldberger's example
has also endured also because there are so few real-life examples
of equity-promoting behavioral interventions to draw on—few, but
not zero. Particularly in recent years, with the advent of polygenic
scores created from well-powered GWAS, we have finally begun to
see examples of enriched environments that both improved outcomes, on average, *and* narrowed genetically associated inequalities,
by differentially benefitting those most at risk for poor outcomes.

One example is a study that examined the health consequences
of an educational reform that happened in the UK in the middle of
the twentieth century: Everyone born on or after September 1, 1957
had to stay in school until their sixteenth birthday.[30] People who
were born right before this cut-off are not expected to differ in any
systematic way from people who were born right after that cut-off.
Birthday, then, forms the basis of a type of natural experiment testing the effects of being forced by the government to stay in school
an extra year.

On average, extra education improved people's health: people who experienced the educational reform had smaller body mass index and better lung function in adulthood. But not everyone responded equally. The effects of the school reform were largest for people who had the highest genetic propensity to be overweight, as measured by a polygenic index from a GWAS of obesity. Because it had the biggest effects for the most at-risk people, the educational reform narrowed genetically associated inequalities. For people who didn't experience the reform, the one-third of people who had the highest genetic risk had a 20 percent greater risk of being over-weight or obese than those who had the lowest genetic risk. After the reform, that gap shrank to just 6 percent.

Another example is an intervention program, called The Family Check-Up, that teaches the parents of teenagers what are called "family management" strategies—how to monitor your teenager's friends and whereabouts without being too intrusive, how to set reasonable limits (like curfew) and enforce them.[31] A randomized controlled trial in the US compared families who had received the intervention to a control group and found that, on average, the inter-vention reduced teenage drinking and the subsequent development of alcohol problems. A follow-up study introduced genetic data, using a polygenic index created from a GWAS of alcohol depen-dence. Among people whose families had been in the control group, the association between the polygenic index and alcohol problems was what you'd expect based on the original GWAS: people with a higher polygenic index had more alcohol problems, on average. But among people whose families had received the intervention, the genetic effect was switched off—there was no association between genetic risk and alcohol problems.

Finally, we see a similar pattern of floor-raising in our study of polygenic associations with mathematics tracking and persistence in US high schools, which I told you about in the last chapter. Overall, students with higher polygenic indices took math courses for more years, on average. (This study used data from the mid-1990s, when only two or three years of math were required in most US states.) But students from advantaged high schools, where more of the families had college degrees, were buffered from dropping out of math, even if they had low polygenic indices.

It's not clear yet why this is. Advantaged high schools could offer additional tutoring or mentorship to struggling students, or there might just be strong social norms among college-educated families about what types of math classes students are expected to take. But two students who have equivalent polygenic indices show different outcomes depending on their school context, particularly if they had low polygenic indices.

These three examples illustrate a more general point: that equity and quality are not always in opposition. In all three examples, a positive environmental difference disproportionately improved the outcomes of people who were at highest genetic risk for poor outcomes—for obesity, for alcohol problems, for math drop-out—and therefore equalized the outcomes of people across genotypes.

Left Behind: When the Rich Get Richer

There is, however, nothing inevitable about equity-promoting outcomes. Interventions can be successful in the sense that they improve outcomes on average, but they can at the same time exacerbate genetic differences between people. For instance, taxing cigarettes and other tobacco products has successfully halved tobacco use since the 1960s. But the health economist Jason Fletcher and others have suggested that taxation has been most effective at discouraging smoking among those who are least genetically at risk for addiction to cigarettes.[32] Those who are most genetically at risk, in contrast, have been increasingly left behind, continuing to struggle with the devastating health consequences (and punitive economic costs) of tobacco use.[33]

This situation, in which the previously advantaged benefit the most from a policy or intervention, is referred to as a Matthew effect, so-called because of Jesus's words in the Gospel of Matthew (25:29): "For whoever has will be given more, and they will have an abundance. Whoever does not have, even what they have will be taken from them." Educational researchers have examined Matthew effects in relation to factors like children's previous test scores or their socioeconomic status, and have found that these effects, while not inevitable, are pervasive when interventions or programs are universally accessible.[34] Summer school programs, for instance, tend to benefit children from middle-class families more than ones from poor families.[35]

Who Is Being Served? A Call for Greater Transparency

Thus far in this chapter, we've discussed three ways that society could be different than it currently is. One, it could become more repressive and impoverished, minimizing inequality of outcome by leveling everyone down. Two, it could minimize inequality of outcome by differentially investing in those most genetically at risk for poor outcomes. Or, three, it could implement interventions and programs that maximize the outcomes of those who are already advantaged but fail to boost (or fail to boost as much) the outcomes of others.

The first alternative possible world, which makes everyone worse off and no one better off, is clearly not preferable. But the choice between the other two alternatives might not be so clear-cut.

In figure 8.3, I've illustrated these alternative environments. For each alternative, there is a distribution of possible outcomes across individuals, but the distributions differ not only in their *average* but also in their *range* (i.e., in how wide the inequalities are). The figure also pinpoints the outcomes of two hypothetical individuals who have different genes: genotype A (circle) versus B (triangle). How the expected phenotypic outcomes of individuals with a certain genotype vary across alternative environments is the *reaction norm.*[36]

The idea of shifts in reaction norm emphasizes a different question than how individuals differ from other people in the same society. Instead of comparing people to each other, we are back to the "What if?" question that I emphasized at the beginning of this chapter: What if your genotype were exactly the same, but the social context changed? In other words, we are now comparing each person to themselves across alternative possible worlds, rather than comparing people to each other within a world. Considered this way, the salient question is not which world minimizes the inequalities in

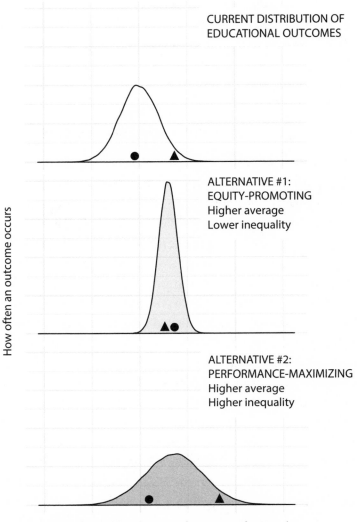

CURRENT DISTRIBUTION OF
EDUCATIONAL OUTCOMES

ALTERNATIVE #1:
EQUITY-PROMOTING
Higher average
Lower inequality

ALTERNATIVE #2:
PERFORMANCE-MAXIMIZING
Higher average
Higher inequality

How often an outcome occurs

Range of possible educational outcomes for people
with different genotypes

FIGURE 8.3. Distribution of educational outcomes for people with different genotypes in alternative environments. The circle and triangle represent two hypothetical individuals with two different genotypes. Relative to the current situation, the environment that is equity-promoting (alternative #1) improves the educational outcome of the individual represented by the circle, but makes little difference for the individual represented by the triangle, reducing inequality of outcome. In contrast, the environment that is performance-maximizing (alternative #2) improves the educational outcome of the individual represented by the triangle but not the individual represented by the circle, thus increasing the inequality of outcome between them but also leading to the highest individual outcome achieved across alternatives.

outcome among children, but which world maximizes the outcome for an individual person.

But whose outcomes do we prioritize? Consider, for instance, a school that is adopting a new math curriculum and has a choice between two proposals. Which would be more preferable—(a) or (b)?

(a) The new curriculum is particularly helpful for children who are most genetically "at risk," thus reducing the gap in educational outcomes between children who did and who did not happen to inherit a particular combination of genetic variants.

(b) The new curriculum is particularly helpful for children who are most likely to succeed anyway, thus inculcating even higher levels of mathematics skill among a few students.

Reasonable people could make a variety of empirical arguments for (a) versus (b). For instance, one might bring various cost-benefit analyses to bear: How many students are helped by (a) versus (b)? How much will a new curriculum cost per student? What are the downstream impacts (in terms of economic productivity, technological innovation, social cohesion, political participation, etc.) of having more people in a society who have a certain baseline level of mathematical skills versus having more people in a society who have a very high level of mathematical skills?

In addition to these empirical questions, however, this choice also involves questions about people's *values*, including whether one values equality of educational outcomes as an end, a good thing to be pursued for its own sake, or simply as a means to some other goal, such as equality of economic outcomes.

Currently, however, policymakers and educators do not have to be transparent about those values, nor do they have to be confronted with evidence regarding whether the realized effects of policies or interventions are living up to those values. In educational and policy research, genetic differences between people are largely invisible, because researchers do not even try to measure anything about people's genetics. When it comes to studying educational attainment or psychological health, well-done studies on gene-by-intervention interactions, like the one that showed how an educational reform in the UK particularly benefitted people at high genetic risk for obesity, are rare to the point of nonexistent.[37]

In my experience talking with scientists who are developing and testing interventions, they are often reluctant to add genetic information to their research. They have pragmatic objections: "Will it be expensive? Will asking for DNA drive down participation? Do I have to have a low-temperature freezer on site?" (The answers are: No, no, and no. Genotyping can be done relatively inexpensively, costing less than $75 per person. As the success of direct-to-consumer genotyping companies attest, people can be quite curious and enthusiastic about participating in genetic research. And saliva samples can be stored at room temperature for months, or even years.)

Underneath these pragmatic objections, however, lurk deeper fears—that the mere act of *collecting* DNA in studies of educational or mental health outcomes inherently buys into a eugenic ideology that the reluctant researcher finds odious, that the risks of letting the genetic genie out of the bottle always outweigh any benefits. Researchers are correct, of course, that there are potential misuses of genetic information, such as using polygenic indices to select individual students for competitive academic positions. I will return to these misuses, and their anti-eugenic alternatives, in chapter 12.

But, like an ostrich sticking its head in the sand, blinding ourselves to genetic data doesn't make genetic differences go away. An intervention that leaves the children most genetically at risk for poor educational outcomes even further behind does so whether that inequality-promoting effect is observed by researchers or not. Rather, ignoring genetics deprives us of another tool for seeing who is being served—and who is not—by our existing roster of interventions.

Equity of What? Remembering the Long Causal Chain

Questions about whether educational interventions or policies *are* equity-promoting, and about whether they *should* be, are weighted with extra significance because differences between people in their educational attainment are so closely tied to many other forms of inequality. Particularly in the United States, disparities in income, wealth, physical health, and psychological well-being between those who have and those who have not obtained a college degree are wide and getting wider (as I described in chapter 1). Too often, increasing equity in educational outcomes, by increasing the number of people who get through college and making college less dependent on the circumstances of one's birth, whether genetic or socioeconomic, is portrayed as the only possible means for addressing this broader set of inequalities in people's lives.

But narrowing gaps in educational outcomes is not our only recourse for addressing the economic and health crises that face Americans without a college education. Certainly, I believe that education is a good to be pursued for its own sake, and that providing more people with the real opportunity to spend several years studying arts or literature or science or philosophy would enrich their lives. We can maintain that education is a good thing for people to acquire, however, without fetishizing higher education as the only acceptable pathway to building a healthy, secure, and satisfying life. As the economists Anne Case and Angus Deaton wrote, "We do not accept the basic premise that people are useless to the economy unless they have a bachelor's degree. And we certainly do not think that those who do not get one should be somehow disrespected or treated as second-class citizens."[38]

As I described at the beginning of this chapter, hereditarian pessimism about the prospect of social change through social policy has typified eugenic thinkers for over a century. This pessimism grows out of a flawed genetic determinism, which imagines that people's characteristics—their cognition, their personality, their behavior—are inexorably fixed by DNA. Genetic determinism is false, as the myriad studies that I've described in this chapter show, but genetic determinism is not the only falsehood that needs to be uprooted. Hereditarian pessimism about the prospect of social change through social policy also grows out of a flawed *economic* determinism, which imagines that people who don't succeed in education must be inexorably consigned to bad jobs, low wages, and poor health care (or none at all).

We don't have to look very far, however, to imagine different ways of arranging society. Case and Deaton, for instance, argue that much of the blame for the immiseration of non-college-educated

Americans can be laid at the doorstep of our exorbitantly costly health care system, a system that is an outlier among high-income countries.[39] I argued in Part I of this book that genetic differences between people cause differences between them in their social and behavioral outcomes, but that genetic causation must be understood in terms of a lengthy and complex causal chain that spans multiple levels of analysis, from the actions of molecules to the actions of societies. The length and complexity of this causal chain means that there are multiple opportunities to intervene in the connection between genotype and a complex phenotype. Changing the health care system so that wages for "low-skilled" workers were not dragged down by the immense cost of employer-provided health insurance would not change anything about people's DNA—but it might weaken one link in long causal chain connecting genetic differences among people with differences in their income.

Moreover, the relative emphasis we put on equity can differ at different links in the chain. For example, one might readily conclude that an effort to use gene editing to equalize people with regard to their DNA sequence itself would be ghastly in its invasiveness and expense and risk for negative outcomes. One might decide equalizing people with regard to their likelihood of obtaining a PhD in a STEM discipline is less important than maximizing the productivity of a few people with high levels of mathematics interest and ability, even if those interests and abilities are the products of "winning" the social or natural lottery. Yet one might also decide it is important to equalize people with regard to their access to clean water and nutritious food and health care and freedom from physical pain, regardless of their level of education. The long causal chain connecting the genetic lottery with social inequality means that decisions about equity—about what we want the world to look like—must be made at every link in the chain.

Hoping for a Different Kind of Human Society

For many people, one of the great obstacles to accepting links between biology and social behavior is the idea that biology provides a hard stop against the possibility of progressive social change.

This is not an accident. As I discussed in the beginning of this chapter, political extremists have spent much of the past century declaring this to be true. If it's heritable, their reasoning goes, it can't be changed, and so there is no sense in trying to change it, or in imagining "a very different kind of human society," as the philosopher Peter Singer put it.[40] Here, I've begun to describe why genetic causes are not an enemy of social change. But the egalitarian case for acknowledging the importance of genetic causes and embracing the use of genetic research tools doesn't end there.

Using Nature to
Understand Nurture

I argued in the last chapter that the existence of genetic causes of social inequalities does not imply any hard boundaries on the possibility of change. On the contrary, we can find plentiful evidence that environmental changes—ranging from sweeping political changes like the fall of the Soviet Union, to intimate personal changes like family therapy—can change the relationship between people's DNA and their life outcomes. The long causal chain that connects genetics to social inequality might frustrate the scientist, who must contend with its often-baffling complexity, but it offers the parent or policy-maker a lot of different opportunities to intervene.

Given that genetic influence does not operate as a hard upper bound on the possibility of social change, it might be tempting to conclude that those interested in social change can safely ignore genetics. Certainly, many of my academic colleagues think that the field of behavior genetics is, at best, irrelevant to what they do and, at worst, is a pernicious distraction from the work of finding social causes of social phenomena. This, however, is a mistake. In this chapter, I will explain why genetics, far from being an enemy to, or a distraction from, the effort to improve human lives, is in fact an essential ally, one that is abandoned only at great cost.

We Don't Already Know What to Do

Reflecting this belief that genetics only distracts from understanding the "real" causes of social inequality, the bioethicist Erik Parens bemoaned the amount of research funding devoted to genetics in *Scientific American*: "We continue to overinvest our hope in genetics . . . The tools of genetics research . . . will not reduce, much less eliminate, the health disparities that are produced by the unjust social conditions."[1]

Those who, like Parens, see genetics as an overhyped distraction from addressing the social determinants of inequality often assert that the insights and tools of genetics are unnecessary *because we already know what to do* to address inequalities in education, health, and wealth. The educator John Warner, for instance, wrote a response to my work in *Inside Higher Education* arguing that genetic data was not just distracting but dangerous.[2] According to Warner, he "cannot imagine a subject on which we know more about [*sic*] than the environments under which children learn best. . . . We know what to do for students. . . . It's not mysterious."

Building on Warner's argument, the sociologist Ruha Benjamin similarly protested in her book, *Race After Technology*,[3] that the problem facing those who would improve children's lives is "not a lack of knowledge!" She continued, "It is not the facts that elude us, but a fierce commitment to justice." In her view, genetic researchers who want to incorporate new sources of data to study the environment are participating in the "*datafication of injustice*—in which the hunt for more and more data is a barrier to acting on what we already know."

Reading such assertions, one might imagine that there is a vast repertoire of policies and interventions that have been proven to be effective at addressing social inequalities in education and health, and that are just waiting in the wings to be deployed, if we can only muster sufficient political will. But, in fact, experts in the fields of education, behavioral intervention, and social policy have repeatedly reminded us that, often, well-intentioned efforts to improve people's lives fail to make any difference at all, and sometimes make things worse.

In the world of education, one can glimpse the paucity of successful intervention research by perusing the What Works Clearinghouse,[4] a resource curated by the Institute of Education Sciences, the research and evaluation arm of the US Department of Education. A review of randomized controlled trials (RCTs) conducted by the IES concluded: "A clear pattern of findings in these IES studies is that the large majority of interventions evaluated produced weak or no positive effects compared to usual school practices."[5] Similarly, a 2019 review of 141 RCTs in the US and the UK found that their average effect size was less than one-tenth of one standard deviation (.06 SDs). Reckoning with this track record, the authors suggested one possible explanation: "The basic research on which educational interventions are based is unreliable. . . . Interventions that are based on insights gained from unreliable basic research are unlikely to be effective even if they are well designed, successfully implemented, and appropriately trialed."[6]

Similarly, a report by the Laura and John Arnold Foundation (now Arnold Ventures), a philanthropic organization dedicated to finding "evidence-based solutions" for social problems, summarized, "Studies have identified a few interventions that are truly effective . . . but these are exceptions that have emerged from testing a much larger pool. Most, including those thought promising based on initial studies, are found to produce small or no effects."[7] David Yeager, an intervention researcher and one of my faculty colleagues at UT, put it this way: "Nearly all past high school programs—tutoring programs, school redesigns, and more—showed no significant benefits on objective outcomes."[8]

The conclusion either that most interventions don't work, or that no one has ever even studied whether they work, extends beyond academic performance. The developmental psychologist Larry Steinberg reviewed the effects of school-based intervention

programs designed to reduce teenagers' alcohol and drug use, con-
domless sex, and other behavioral risks. An estimated 90 percent
of American adolescents have been forced to sit through at least
one such program. Steinberg concluded: "Even the best programs
are successful mainly at changing adolescents' knowledge but not
in altering their behavior." He went on to note that failure isn't free:
"Most taxpayers would be surprised—and rightly angry—to learn
that vast expenditures of their dollars are invested in . . . programs
that either do not work . . . or are, at best, of unproven or unstudied
effectiveness."[9]

These sorts of conclusions, from interventionists who really and
truly want to make a positive difference in the world, should be
humbling. They should make us think twice before asserting that
we already know what to do to improve people's lives. They should
make us realize that a lack of knowledge and a paucity of data really
are parts of the problem. And, they should remind us that understanding human behavior, much less intervening to change it, is a
hard problem to solve.

Why the Social Sciences are the Hardest Sciences

The psychologist Sanjay Srivastava has a blog called "The Hardest Science."[10] The title is a play on words. Natural sciences (like
physics, chemistry, and biology) are typically deemed the "hard"
sciences, considered purer and more rigorous than the so-called
"soft" sciences (like psychology, sociology, economics, and political
science) that study the functioning of human society and the behavior of individuals within a society. As an editorial in *Nature* put it,
"Soft . . . too readily translates as meaning woolly or soft-headed,"
when, in fact, "the social sciences are among the most difficult of
disciplines, both methodologically and intellectually."[11] By naming
his blog on research in psychology "The Hardest Science," Srivastava was calling attention both to the methodological features that
psychology shares with the so-called "hard" sciences (e.g., using
controlled experiments) and to the fact that psychology—like the
other social sciences—focuses on hard problems. Why, for instance,

do some children learn what is taught in school so much more easily than other children, and what should we change in order to help children who struggle in school? The too-common failure of promising educational interventions to deliver real change shows us that these are not easy questions with easy answers.

In particular, Srivastava points to three features that make psychological problems hard problems to solve. First, human behavior is embedded in complex systems at multiple levels of analysis. The brain is a complex system, and so is society. Often, we are interested in isolating the effects of one feature of that complex system—if I changed x and only x, what would happen? This is challenging even when people can do randomized experiments (like the Romanian orphanage study that I described in chapter 5), but is made dramatically more complicated when it is ethically or pragmatically impossible to do an experiment. (We will return shortly to an example where experimentation is essentially impossible.) Second, unlike natural laws that are true at all places and times, the rules that govern the functioning of societies and human behavior within societies vary by local conditions. (Indeed, the inability of psychologists and other social scientists to come up with causal rules that are true at all times and places has been a major source of exasperation on the part of biologists.) And finally, human psychology and behavior involve concepts that are difficult to quantify. What is the appropriate scale for measuring happiness? Satisfaction with life? Intelligence?

Because human behavior and social structures are complex, doing good social science research would be hard even if people did not differ genetically in ways that matter for their lives. But they do. And the ubiquity of genetic differences relevant for socially important traits makes the "hard" science of psychology even harder.

Recall the first reason why, Srivastava argued, psychology has hard problems: human behavior is embedded in a complex system of interacting parts, whereas we are often interested in isolating one part and understanding what will happen if we change it. What happens, for instance, if parents of young children talk more to them in the first three years of their life—but nothing else changes? Will pulling on that one lever of environmental change make a positive difference for children's lives—their cognitive development and performance in school, for instance? Answering that question is hard, because parents who talk a lot to their young children might differ, in lots of other ways, from parents who don't. They might be richer. They might have more-regular work schedules. They might send their children to different sorts of preschools. And, they might have different DNA, which is inherited by their children. Because these different factors are braided together, parents talking more to their young children might be *correlated* with how well those children do in school, but that doesn't mean that *changing* how much parents talk to their children will make a difference in how well those children do in school. We are back to the idea that correlation does not equal causation.

The idea that genetic differences between people are braided together with the environmental differences that social scientists seek to understand and change can be met with hostility. When I wrote in the *New York Times* that genetic research related to education would be helpful for understanding environmental levers for change,[12] the sociologist Ruha Benjamin accused me of engaging in "savvy slippage between genetic and environmental factors that would make the founders of eugenics proud."[13] But the "slippage" between genetic and environmental factors is not an invention of eugenic ideology. It is, rather, a byproduct of the fact that humans exist at the boundary between the natural and the social worlds. That genetic and environmental factors are braided together is simply a description of reality.

A century of the so-called "nature-nurture debate" has conditioned people to think of genes as competing against the environment in a zero-sum game, where any attention to biology must necessarily be accompanied by a reduction in the attention paid to society. But to design social interventions and policies to improve people's lives is to ask, "What would happen if x—but only x—changed about people's environments?" Answering such questions requires reckoning with all of the other features of people's lives that ordinarily go along with x—a very long list, a list that includes their DNA.

A Sexy Example

A concrete example might be helpful. How do environmental and genetic factors slip together in the course of ordinary human development, and how might genetic research help us understand the impact of *environmental* experiences?

Here in my home state of Texas, our Education Code mandates that "any course materials relating to human sexuality" must "emphasize that abstinence from sexual activity, if used consistently and correctly, is the only method that is 100 percent effective in preventing . . . the emotional trauma associated with adolescent sexual activity."[14] That's right—Texas students are legally required to learn that having sex as an unmarried teenager causes emotional trauma.

At first glance, the developmental psychology literature on this topic seems to back up the state's claim. Adolescents who have sex at younger ages don't go as far in school, report more psychological distress and higher rates of depression, are more likely to use alcohol and other drugs, are more likely to engage in delinquent and criminal behavior, and, in girls, are more likely to have patterns of disordered eating.[15] On average, earlier age at first sex is *correlated* with generally worse outcomes for teenagers. On the basis of this correlation, the state of Texas has leapt to a causal conclusion: every public school in Texas is required, by law, to teach teenagers that sex *causes* depression and other mental health problems, and that abstaining from sex will prevent these bad things from happening to them.

There are problems, of course, with leaping from correlation to causation. Teenagers who have sex at fourteen are *different* from those who are still virgins at twenty-two, in lots of ways other than their sexual experiences. Here, we see the "slippage between genetic and environmental factors" that characterizes humans. A sexual encounter is a social *environment* that could potentially have a causal impact on your subsequent development. (When did you lose your virginity? Would your life be different if you had waited longer or had sex earlier?) At the same time, the initiation of sexual intercourse is part of a years-long developmental process that transforms a child to a reproductively mature adult, and the timing and pace of reproductive maturation are influenced by heaps of other factors—including genes. And the same genes that predispose someone to have earlier or faster reproductive development could also predispose them toward mental health problems. For instance, one huge study of both men and women in the UK Biobank found that genes associated with earlier age at first sex also conferred risk for ADHD and smoking.[16]

How then should be we interpret the observation that teenagers who have sex earlier are more likely to show emotional and behavioral problems? For the sake of simplicity, let's focus on just two alternative explanations. One, the experience of sex could causally impact later psychological development. Or, two, genes that accelerate reproductive development in teenagers could also confer risk for mental health problems.

As I discussed in chapter 5, one way of testing a causal hypothesis is a randomized controlled trial. Except, we can't directly randomize when a teenager has sex. ("Hi, we've drawn your name out of a hat so now you have to wait until you are 25 to lose your virginity.") We *could* randomize teenagers to sex-education programs that are designed to promote abstinence from all sexual activity, but there is no evidence that those programs, despite the over $2 billion in federal funding that the US has spent on developing and disseminating them, actually do anything to change teenagers' sexual behavior.[17] Or, we could test the experiment in laboratory animals, where we do have experimental control over age at first sex, but it's dicey whether any of the results could be expected to generalize to humans. (Can rats ghost each other?) This challenging situation is all too familiar for social scientists: We have a causal hypothesis we want to test ("adolescent sex causes emotional trauma") but no obvious way to do an experiment to test it.

In one of the first studies I did as a graduate student, I tried to get at this problem by using twin data.[18] Identical twins are matched for their genes *and* for many of the environmental variables (e.g., neighborhood poverty, parental attitudes toward sex, percentage of sexually active classmates, distance from nearest sexual health care provider) that might also contribute to both sexual behavior and risk for mental health problems. The key question is: Do identical twins who *differ* in their age at first sex also show differences in their risk for psychopathology? If earlier sex is an environmental experience that *causes* psychopathology, as Texas's sex education policy claims, then the twin who had sex younger than the other twin should have a higher average risk of psychopathology. If, however, earlier sex is a phenotypic marker of a set of *genetic* risks, then identical twins, who are genetically identical, should show the same risk for psychopathology, regardless of who had sex when.

In a series of studies conducted in the US, Sweden, and Australia, my colleagues and I asked this exact question.[19] That is, we tested whether identical twins who differed in their age at first sex differed in their subsequent outcomes. The answer was generally . . . nope. The correlations between early age at first sex and substance use, depression, criminal conviction, conduct disorder, delinquency, and risky sexual behavior in adulthood *all* disappeared when researchers controlled for genetic differences between people by comparing identical twins. The best explanation for this pattern of findings is that age at first sexual intercourse, while correlated with adolescent mental health and behavioral problems, did not *cause* those problems.

This analysis illustrates three more-general points. First, environmental experiences—whether it be having a sexual relationship at a certain point in one's adolescence or receiving a certain type of parenting or living in a certain type of neighborhood—can be *correlated* with life outcomes but not be *causes* of them.

Second, policies that are built on a flawed understanding of which environments are truly causal are wasteful and potentially harmful. In this specific example, even if the state of Texas was successful at delaying teenagers' sexual activity, such a change would not actually improve their mental health—and an emphasis on such programs potentially diverts investment away from educational programs that *would* be helpful. (Proponents of teenage abstinence might argue that abstinence is a valuable end for its own sake, but that is a different justification for the policy than the empirical claim that abstinence is a means toward increased adolescent well-being.)

Third, genetic data—whether it be comparisons of identical twins or comparisons of people with similar polygenic indices— help researchers solve the first problem and, in so doing, avoid the second problem. Genetic data gets one source of human differences *out of the way*, so that the environment is easier to see.

Getting It Wrong Isn't Free

Of course, it's not just sex education policy that is based on a fast-and-loose reading of correlations that might not be giving an accurate picture of what actually causes what. Consider, for example, the famed "word gap," which is the estimated difference in the number of words that poor children hear before the age of 3, compared to children from high-income families. The original word gap study, based on a sample of 42 families who were recorded for about an hour a week for several years, concluded that poor children heard 30 million fewer words by the age of 3 than affluent children.[20]

The word gap became a darling of academics and policymakers alike. In 2013, the Clinton Foundation announced a public action campaign, which included advertisements filmed by movie-star moms, focusing on closing the word gap: "Poverty of vocabulary should be discussed with the same passion as child hunger."[21] President Barack Obama followed suit in 2014. Citing the "30 million words" figure, he declared that closing the word gap was one of his "top priorities," a goal that was necessary "if we're truly going to restore our country's promise of opportunity for all."[22] Around the same time, the "Providence Talks" program was launched, backed by a multimillion-dollar grant from Bloomberg Philanthropies.[23] The Providence, Rhode Island, program gives participating parents audiometers that track how much they are talking to their kids, and coaching to help them increase their child-directed speech.

With the word gap, people have not asked for more data before diving in to act on what we think we already know. The problem is that no one can agree on what we already know. Nearly every aspect of the conclusions of the word gap study, and whether they should be acted on, is scientifically controversial. Some people argue that the word gap doesn't actually exist, as only some studies have been able to replicate the results of the original study.[24] Other people don't disagree that there *is* a difference in the number of words that children from different groups hear, but they object to using the word "gap." Why do we assume that the linguistic norms typical of middle-class White people in the United States are the standard to which everyone else should aspire? Maybe this is just another way in which we unfairly stigmatize poor families as deficient, rather than just culturally different.

But there is also a glaring problem that few people address at all: parents are genetically related to their children. And the same genes that are associated with adults' educational attainment and income and occupational status are also associated with how early children begin to talk and how well they read at age seven[25]—the same vocabulary outcomes that are allegedly the outcome of being exposed to more child-directed speech. There is a striking paucity of early language research that even nods at a potential role of genetics in explaining why parents who talk more have children who talk more.

The jury is still out on whether "word gap" interventions will be effective. But the *premise* of the word gap intervention is terribly shaky—observing a parent-child correlation and assuming that it represents a causal effect of the parentally provided environment. What if that premise is wrong? Before we spend literally millions of dollars on interventions designed to change a parental behavior in hopes of improving child outcomes, I believe it would be prudent to at least *check* to see that the correlation between parental behavior and child outcome is still there when we control for the fact that parents and children share genes. For instance, do adoptive parents who say more words to their children have adopted children who have better earlier reading? If not, this result would cast serious doubt on the idea that the number of words heard is *causing* differences in child literacy outcomes—and serious doubt on the idea that word gap interventions will ultimately prove to be effective at improving children's outcomes.

Every policy decision involves trade-offs: Investing in one thing, like the word gap, necessarily involves not spending that time and money on something else, which could have been more effective at producing one's desired ends. Ultimately, *all* interventions and policies are built on a model about how the world works: "If I change x, then y will happen." A model of the world that pretends all people are genetically the same, or that the only thing that people inherit from their parents is their environment, is a wrong model of how the world works. The more often our models of the world are wrong, the more often we will fail in designing interventions and policies that do what they intend to do, and the more often we will face the unintended consequences of not investing in something more effective.

The "Tacit Collusion" to Ignore Genetics

Disappointingly, rather than addressing this problem, many scientists in the fields of education, psychology, and sociology simply pretend it doesn't apply to them. The sociologist Jeremy Freese summarized the situation as follows:

> Currently, many quarters of social science still practice a kind of epistemological tacit collusion, in which genetic confounding potentially poses significant problems for inference but investigators do not address it in their own work or raise it in evaluating the work of others. Such practice involves wishful assumptions if our world is one in which "everything is heritable."[26]

Freese was writing in 2008, but the situation now is no different. Open almost any issue of a scientific journal in education or developmental psychology or sociology, and you will find paper after paper announcing correlations between parental characteristics and child development outcomes. Parental income and child brain structure. Maternal depression and child intelligence. Each of these papers represents a massive amount of investigator time and public investment in the research process, and each of these papers has, in Freese's words, an "incisive, significant, and easily explained flaw"—that differences in children's environments are entangled with the genetic differences between them, but no serious effort is being expended toward disentangling them.

The tacit collusion among many social scientists to ignore genetics is motivated, I believe, by well-intentioned but ultimately misguided fears—the fear that even *considering* the possibility of genetic influence implies a biodeterminism or genetic reductionism they would find abhorrent, the fear that genetic data will inexorably be misused to classify people in ways that strip them of rights and opportunities. Certainly, there are misuses of genetic data that need to be guarded against, which I will return to in chapter 12. But while researchers might have good intentions, the widespread practice of ignoring genetics in social science research has significant costs.

In the past few years, the field of psychology has been rocked by a "replication crisis," in which it has become clear that many of the field's splashy findings, published in the top journals, could not be reproduced and are likely to be false. Writing about the methodological practices that led to the mass production of illusory findings (practices known as "p-hacking"), the psychologist Joseph Simmons and his colleagues wrote that "everyone knew [p-hacking] was wrong, but they thought it was wrong the way it is wrong to jaywalk." Really, however, "it was wrong the way it is wrong to rob a bank."[27]

Like p-hacking, the tacit collusion in some areas of the social science to ignore genetic differences between people is not wrong in the way that jaywalking is wrong. Researchers are not taking a victimless shortcut by ignoring something (genetics) that is only marginally relevant to their work. It's wrong in the way that robbing banks is wrong. It's *stealing*. It's stealing people's time when researchers work to churn out critically flawed scientific papers, and other researchers chase false leads that will go nowhere. It's stealing people's money when taxpayers and private foundations support policies premised on the shakiest of causal foundations. Failing to take genetics seriously is a scientific practice that pervasively undermines our stated goal of understanding society so that we can improve it.

There is another danger, too. To return to Freese's assessment of social science that ignores genetics: "While particular areas might be quite productive—yielding literature that can be summarized as saying 'many studies show x'—they are chronically vulnerable to sweeping dismissal from outside."[28] Freese was concerned about sweeping dismissals from other academics outside his home field of sociology. But I am more concerned about sweeping dismissals of social science research from political extremists.

When social scientists routinely fail to integrate genetics into their models of human development, they leave space for a false narrative that portrays the insights of genetics as a Pandora's box of "forbidden knowledge."[29] That is, social scientists are increasingly portrayed as deliberately censoring or "canceling" any study of genetic differences, because the data will inevitably prove, in this line of thinking, the deterministic idea that a person's life outcome is only a product of their DNA.

Moreover, people who are invested in maintaining the status quo of social inequality can easily critique studies purporting to show the negative impacts of environmental conditions, pointing out that far too many do not rigorously control for the fact that people's environments are entangled with genetic differences between them. Why would we want to hand people opposed to the goals of social equality a powerful rhetorical weapon, in the form of a widely prevalent and easily understood methodological flaw in social research? In contrast, when social scientists take genetics seriously, they can demonstrate the negative impacts of environmental conditions all the more clearly.

New Tools for Old Problems

It is worth noting again that no serious scholar thinks that inequality is entirely "genetic." The lesson of the research I told you about in the first half of this book is *not* that the Environment, writ large, doesn't make a difference for people's lives. Rather, one lesson is that figuring out which specific environments make a difference, for whom, at what point in the life span, is a harder problem than it might appear at first, because most of those environments are braided together with genetic differences between people.

Addressing this challenge is why many researchers are excited about twin studies, adoption studies, GWAS, and polygenic indexes. Researchers want tools to make genetics recede into the background, *to get it out of the way*. This motivation for doing genetic research is readily apparent when I talk to my colleagues. Often, the phrase that animates them the most is not a phrase with clickbait allure like "embryo selection" or "personalized education." Rather, the phrase scientists who do work in this area keep returning to is a dry-sounding statistical concept—"control variable."

Look, for example, at the extensive FAQ written by the Social Science Genetics Association Consortium to accompany the publication of their 2018 GWAS of educational attainment.[30] It was extremely pessimistic about using an education-associated polygenic index for "*any* practical response," because the index "is not sufficient to assess risk for any specific individual." The only application they did endorse? "The results of our study may be useful to social scientists, e.g., by allowing them to construct polygenic scores that can be used as control variables."

Similarly, Sam Trejo and Ben Domingue, educational researchers from Stanford University, introduced one of their scientific papers talking about the "great promise" of polygenic indices. The reason for this promise? Because they "may be used as control variables in studies of environmental effects."[31] Or look at the website of Dalton Conley, a sociologist at Princeton. Conley avows that he's particularly excited about polygenic indices—because including them as control variables in statistical analyses allows one to "obtain better-specified, less-biased parameter estimates for [environmental] variables."[32]

While the media discusses behavioral genetic research in terms of designer babies and surveillance capitalism, the researchers who are creating and working with polygenic indices for traits such as education are geeking out about control variables and less-biased parameter estimates. Discussions of control variables don't have the same dark allure as discussions of designer babies. But much of the potential for genetic research to improve human lives resides here, in thinking about doing better social science research.

One important study that showcases the power of genetic data to study the environment was conducted by Dan Belsky and colleagues. I've told you part of their results in earlier chapters: children who inherit more education-associated genetic variants than their

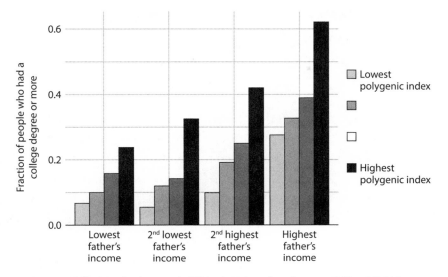

FIGURE 9.1. College graduation rates in White Americans born between 1905 and 1964, by paternal income and by polygenic index created from GWAS of educational attainment. Data courtesy of Nicholas Papageorge and Kevin Thom; results described in Nicholas W. Papageorge and Kevin Thom, "Genes, Education, and Labor Market Outcomes: Evidence from the Health and Retirement Study," NBER Working Paper 25114 (National Bureau of Economic Research, September 2018), https://doi.org/10.3386/w25114.

siblings grow up to be wealthier and to be employed in higher-status occupations than their siblings.[33] This sibling comparison focuses on people who have the same home environment but differ genetically. But we can also look at the people who have the same education-associated polygenic index but who were born into families that differed in their social class. Here, the power of the family environment is spotlighted: children with high polygenic indices but whose parents had the lowest socioeconomic status still ended up, on average, worse off as adults than children who had low polygenic indices but had wealthy parents.

The economists Kevin Thom and Nicholas Papageorge came to a similar conclusion in their analysis of college graduation rates[34]: 27 percent of rich children with the lowest polygenic indices graduated from college, compared with 24 percent of poor children with the highest polygenic indices (figure 9.1).

These types of results about genetics and social mobility are like the face/vase illusion: focusing on one part of the picture makes the other part recede in your perception, but a shift in attention can bring that other half forward again. At every level of the social ladder, children who have a certain constellation of genetic markers are more likely to be upwardly socially mobile than other children who did not inherit those markers. But if they are born into poverty, even the most genetically advantaged children will *still* have a lower socioeconomic status in adulthood than children who have no genetic advantages but were born to wealth. As the social scientist Ben Domingue summarized, "genetics are a useful mechanism for understanding why people from relatively similar backgrounds end up different. . . . But genetics is a poor tool for understanding why people from manifestly different starting points don't end up the same."[35]

Yet another line of research has cleverly used genetic information from children and their parents to spotlight the effects of the environment. Remember that every parent has two copies of every gene, only one of which is transmitted to a child. For every parent-child dyad, then, the parent's genome can be divided into the *transmitted* alleles (those genetic variants the child inherited) and the *untransmitted* alleles (the genetic variants the child did *not* inherit). Essentially, this is dividing the parental genome into one part that is like an adoptive parent's (not resembling the child at all) and another part that is like an identical twin (exactly the same as the child).

The critical test then becomes: Are a parent's *untransmitted* genes nonetheless related to their child's life outcomes?[36] If so, then, there is an association between parental genes and child phenotype that *cannot* be due to genetic inheritance from parent to child; the association must be due to some part of the *environment* provided by the parent.

One of the largest studies using this type of research design was conducted in Iceland, a small country that looms large in genetics research because it is ancestrally homogeneous, has "exquisite" medical and genealogical records, and has genotyped more than one-third of its population.[37] This study found that for physical traits like BMI or height, your parents' genes didn't make you taller or fatter unless you actually inherited them; the untransmitted alleles were uncorrelated with the child phenotype. For education, on the other hand, your parents' genes are still associated with your own ultimate educational attainment—even if you didn't inherit those genes. By ruling out biological inheritance as a mechanism for why parental characteristics were correlated with their children's outcomes, the study showed that it *must* be the environment provided by the parents that was shaping the children's educational trajectories. Taking genetics seriously allowed the effects of the environmental privilege to be seen more crisply, providing a direct rebuttal to the eugenic argument that the apparent social determinants of inequality are "really" just unmeasured genetic differences.

Using Every Tool in the Toolbox

There are practical problems that need to be overcome before genetic data, in the form of polygenic indices, can be more routinely integrated into policy and intervention research. As of this writing, the biggest practical problem is that, as I explained earlier in the book, we do not have polygenic indices that are statistically useful for studying health and achievement outcomes in people who aren't of European genetic ancestry. In the United States, more than half of public-school children have a racial identity that is not White and so can be reasonably expected to have at least some non-European genetic ancestry. The children who are often most in need of improved educational interventions, then, are the same ones for whom we have the fewest tools in the genetic toolbox. The statistical geneticist Alicia Martin summarized the problem this way: "To realize the full and equitable potential of [polygenic indices], we must prioritize greater diversity in genetic studies . . . to ensure that health disparities are not increased for those already most underserved."[38]

This problem has the potential to be solved, however, as genetic research becomes more global. I anticipate that scientists will have developed a polygenic index that is as strongly related, *statistically*, to academic achievement in Black students as it is in White students. In fact, I anticipate that scientists will have made significant progress on this problem long before we will have developed better policy and intervention tools that, say, reliably increase high school graduation rates for teenagers struggling in math, or decrease motor vehicle accidents among teenagers with ADHD. As I described in chapter 4, it would be scientifically and ethically wrong to use such polygenic indices to make comparisons between different racial groups. Moreover, *which* genes are associated with academic achievement, and how strongly those genes are associated with academic achievement, might differ across people who differ in their genetic ancestry. Extending the GWAS revolution to groups other than European ancestry populations would nonetheless allow researchers to conduct studies *within* each group that are similarly rigorous in their ability to identify specific environmental causes of important developmental outcomes.

Given the formidable obstacles to creating interventions and policies that improve people's lives, we shouldn't expect too much of any one research methodology. Certainly, a broader use of genetic data in the social sciences will not solve every problem. But genetic data can *help* in the effort to improve people's lives. As I've already described, all interventions and policies reflect a model of how the world works. If basic research about education or child development is flawed or unreliable, it is harder to design interventions and policies to improve people's outcomes. The biggest contribution of genetics to the social sciences is to give researchers an additional set of tools to do basic research by measuring and statistically controlling for a variable—DNA—that has previously been very difficult to measure and statistically control for. As genetic information gets cheaper and more broadly applicable, I hope we are not still falsely proclaiming that we already know everything we need to know about how to improve children's lives, and instead are prepared to use every tool in our toolbox.

10

Personal Responsibility

"I just pray . . . that they can go to the afterlife and be wherever God want them to be."

Tears were rolling down Amos Wells's face as he talked about the deaths of his twenty-two-year-old pregnant girlfriend, Chanice, along with her ten-year-old brother, Eddie, and her mother, Annette. The night before, Wells had gone to their home in Fort Worth, Texas, and shot them multiple times. He then turned himself in to police. From jail, he gave a seven-minute-long interview to a local NBC reporter, which aired online in 2013.[1]

"There's no explanation that I could give anyone, or anybody could give anyone, to try to make it right, or make it seem rational, to make everybody understand . . . There's no reason."

Despite Wells's protestations that there was no explanation, his defense attorneys sought one in the field of genetics. Drawing on a candidate gene study conducted in New Zealand, the defense argued during Wells's sentencing phase that he had a propensity for violence because he had inherited a certain version of the *MAOA* gene—what one expert witness called "a very bad genetic profile."[2] You can flip back to chapter 2 for a discussion about why scientists are no longer convinced that these sorts of candidate gene studies are trustworthy.

The jury in Mr. Wells's case was also unconvinced. They unanimously voted to sentence him to death.

As a behavioral geneticist, reading the transcript of Mr. Wells's sentencing proceedings is a humbling experience. I know and respect the senior authors of the original candidate gene study linking *MAOA* with criminal behavior; I myself have published numerous articles on the genetics of teenage delinquency. When writing an article for an academic journal, it is difficult for me to imagine that my dry, jargon-filled prose might be quoted by an "expert" in a Texas courtroom while twelve people decide whether the state should kill a man. A nagging question attends all of our science: If you take genetics seriously as a source of differences between people, then what does that mean (if anything) about the responsibility that we bear for how our lives turn out? This question can no longer be waved off as abstract when genetics is entered into evidence for a capital punishment case.

How genetics affects our judgments about personal responsibility is a question that is not limited to the domain of criminal behavior. Throughout this book, I have tried to make the case that genetics should be taken seriously as an accident of birth that influences one's educational trajectories. Just as those who harm others are punished by the state, those who "succeed" in school are rewarded by society. The educated are rewarded not just with more money and more stable employment, but with better health and well-being.

In this chapter, I next consider how the existence of genetic influences on education changes our perceptions of how responsible people are for their success or failure in school—and for everything that comes with it. Then, in light of the tension between genetics and "personal responsibility," I consider how genetic research on socioeconomic outcomes might be used to make the case for *greater* redistribution of resources in society.

Genetics of Crime

Wells's defense team was not the first to blame their client's behavior on his genes. A 2017 review of legal databases found that information on a defendant's *MAOA* genotype was submitted as evidence in eleven criminal cases, most commonly in the sentencing phase or in post-conviction appeals.[3] Original studies of how genetic information was used in criminal trials proposed that it might work like a "double-edged sword," potentially making a defendant seem less morally culpable, but also seem more likely to be a continued danger to society.[4]

Subsequent work, however, has found that genetic information introduced in forensic settings is not so much a double-edged sword as a blunt-edged sword. Both for judges and the general population, whether the crime is serious (murder) or minor (property damage), genetic explanations do not generally change the punishment that people think is appropriate. *Genetic information is dismissed when we want to punish people.*[5]

This dismissal of genetic explanations for criminal behavior should surprise us, because judgments about other sorts of behaviors and life outcomes are more easily swayed by genetic information. On average, a "biogenetic" explanation of psychological problems (like depression, schizophrenia, anxiety, obesity, eating disorders, and sexual difficulties) reduces judgments of blame and responsibility for these problems.[6] And the popularization of biogenetic explanations for sexual orientation contributed to an increase in support for gay rights.[7] Accordingly, several activist communities have embraced the potential value of genetic research in fighting stigma. The National Alliance on Mental Illness, for example, releases fact sheets for each psychiatric diagnosis that prominently list "genetics" and "brain structure" as causes of the disease.[8] Genetics can be an antidote to blame.

The difference in how genetic research figures into ascriptions of blame for being depressed or overweight versus blame for committing a violent crime is not easily attributed to differences in the underlying genetic research. While the argument used in Amos Wells's case about the influence of the *MAOA* gene was a scientifically weak one, the evidence that genes matter for aggression and violence is strong. Serious behavioral problems beginning in childhood, physical aggression, and emotional callousness are all part of a syndrome of antisocial behavior that is already highly heritable (>80%) in childhood.[9]

Beyond high heritability, there is also emerging genetic data relevant to the likelihood of criminal offending. In the largest genetic study related to criminal behavior, my collaborators and I pooled information from GWAS of a variety of impulsive or risky behaviors, such as ADHD symptoms in childhood, having sex with lots of partners, alcohol problems, smoking pot. These behaviors are not always illegal, but they are all more common in people who also commit violent crimes. All together, we pooled information on nearly 1.5 million people and tested whether individual SNPs were associated with an overall propensity for what psychologists called "externalizing," which is a persistent tendency to violate rules and social norms and to struggle with impulse control.

Using the results of our GWAS, we found that those who were high on the externalizing polygenic index, compared to those low on the polygenic index, were more than 4 times more likely be convicted of a felony and almost 3 times more likely to be incarcerated (figure 10.1). They were also more likely to use opioids and other illegal drugs, to have an alcohol use disorder, and to report symptoms of antisocial personality disorder, which is a psychiatric condition characterized by recklessness, deceitfulness, impulsivity, aggressiveness, and lack of remorse.[10]

Again, I should emphasize that our research was focused on differences within groups of people who all share European genetic ancestry and who would likely identify as White. As I described in chapter 4, these genetic associations cannot and should not be used to explain racial inequalities in contact with the police or in incarceration rates.

Even with this caveat, people find research on the genetics of criminal behavior disturbing, and indeed, unpacking the study that I just described in the context of what we know about human development and social contexts could fill another book. But, for now, let's do something simpler. Just note how different it might sound to you if I say that "Genes influence people's impulsivity and risk-taking so they are less to blame for committing a crime," than if I say "Genes influence people's body weight so they are less to blame for being overweight," or "Genes influence people's

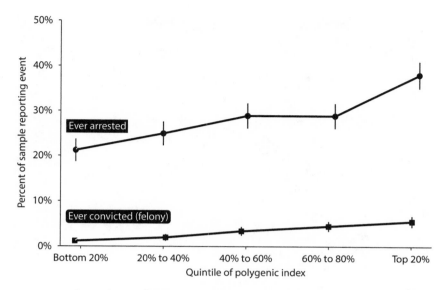

FIGURE 10.1. Rates of criminal justice system involvement and antisocial behavior by polygenic index created from GWAS of externalizing in 1.5 million people. Figure adapted from Richard Karlsson Linnér et al., "Multivariate Genomic Analysis of 1.5 Million People Identifies Genes Related to Addiction, Antisocial Behavior, and Health," bioRxiv, October 16, 2020, https://doi .org/10.1101/2020.10.16.342501.

mood and emotions so they are less to blame for being depressed." Depending on the phenotype, people respond to genetic information differently.

Wanting to Blame

Intrigued by this apparent discrepancy about how information about genetics is received, a psychologist (Matt Lebowitz), a philosopher (Katie Tabb), and a psychiatrist (Paul Appelbaum) teamed up to try to understand why.[11] In a fascinating series of studies, Lebowitz, Tabb, and Appelbaum had their participants read stories about "Jane" or "Tom," who had engaged in either an antisocial behavior (e.g., stealing money from a sleeping homeless man, bullying a younger student) or prosocial behavior (e.g., checking to make sure the homeless man was okay, coming to the defense of a bullied student).[12] Participants were then either given information about Tom

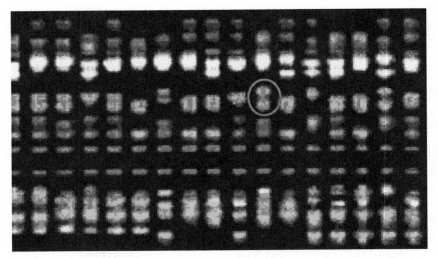

"Scientists have found that people can have genes that lead them to behave this way. Here is a graphic that illustrates the area of the genome where these genes are found. According to recent testing, Jane has these genes. In other words, Jane's genetic makeup—the DNA that she inherited from her parents—leads her to behave the way she does in situations like these."

FIGURE 10.2. Genetic explanation of behavior. Image and text provided to participants in Matthew S. Lebowitz, Kathryn Tabb, and Paul S. Appelbaum, "Asymmetrical Genetic Attributions for Prosocial versus Antisocial Behaviour," *Nature Human Behaviour* 3, no. 9 (September 2019): 940–49, https://doi.org/10.1038/s41562-019-0651-1; image originally from Nicholas Scurich and Paul Appelbaum, "The Blunt-Edged Sword: Genetic Explanations of Misbehavior Neither Mitigate nor Aggravate Punishment," *Journal of Law and the Biosciences* 3, no. 1 (April 2016): 140–57, https://doi.org/10.1093/jlb/lsv053, by permission of Oxford University Press.

or Jane that explained their behavior in terms of genes, or provided no genetic explanation.

The researchers were quite insistent in their presentation of genetic "evidence," providing both a figure and a chunk of explanatory text (figure 10.2).

Participants were then asked how much they believed what they had just been told: *"How much of a role do you think genetics played in Tom/Jane's behaviour in the story you just read?"* (Remember, they have no information about Tom and Jane except what the investigators gave them, and the investigators *just* told them that "Jane's genetic makeup leads her to behave the way she does.") They were also asked, in some studies, *"To what extent to do you believe Jane is responsible for her patterns of behaviour that you just read about?"* and *"To what extent do you think Jane's patterns of behaviour that you just read about reflect who she truly is?"*

Across all the studies, participants were significantly more like to endorse genetic explanations of prosocial behavior than of anti-social behavior. And the tendency to reject genetic explanations for antisocial behavior went along with the tendency to say that Jane was *responsible* for her behavior.

These results suggest that the relationship between our judgments of blame and responsibility and our endorsement of genetic explanations goes in the opposite direction than is commonly assumed. It is not that we hear about genetic influence and decide accordingly whether someone is less responsible. It's that we decide whether or not we want to hold someone responsible and then reject or accept information about genetic influences accordingly.

If we reject genetic evidence when we want to maintain our ability to assign blame and want to hold people morally responsible, this conclusion helps us make sense of results of another study that I previously told you about in chapter 2. In that study, psychologists at the University of Minnesota asked people to estimate how heritable different phenotypes are.[13] The people who were most accurate were mothers with more than one child, but generally folks converged on something resembling the right answer. That is, the average estimate of heritability was reasonably close to the scientific consensus about how heritable a trait is in Western industrialized societies. There were, however, two exceptions. In an interesting twist, people in this study substantially *over*-estimated the heritability of just two phenotypes—breast cancer and sexual orientation. People who described themselves as politically liberal had particularly high estimates of the heritability of sexual orientation.

Breast cancer and sexual orientation are very different types of outcomes, with very different genetic architectures, but they have at least one thing in common: it is generally taboo to imply that someone had a choice about whether or not to get cancer or whether or not to be gay. Telling a victim of breast cancer that they are responsible for their illness is considered victim-blaming. And particularly among political liberals, telling someone who is gay that they are responsible for not being straight is considered homophobic. The flip side of rejecting genetic information to preserve our ability to lay blame is to embrace genetic information in order to deflect blame.

Identical Twins and the Free Will Coefficient

The fact that people's judgments about responsibility go hand in hand with their embrace or rejection of information about heritability raises a more fundamental question—*should* we consider genetic influences as a limitation a person's agency or control? Should we hold people less responsible for their highly heritable outcomes?

This type of discussion has the danger of falling down a black hole of never-ending metaphysical debate. Whether the universe is deterministic, whether such a thing as free will actually exists—these questions are beyond the scope of this book, to put it mildly. We need to put some philosophical guardrails up. If you think that the universe is deterministic, and the existence of free will is incompatible with a deterministic universe, and free will is an illusion, then genetics doesn't have anything to add to the conversation.[14] Genetics is just a tiny corner of the universe where we have worked out a little bit of the larger deterministic chain.

But putting aside metaphysical questions, we, as social beings who live in community with one another, do not treat murder and eye color the same way. In the ordinary course of human affairs, we don't judge people as choosing to be blue-eyed, or as being responsible for making sure they are not brown-eyed, or as deserving more because they are blue-eyed, because we have (rightly, in my view) decided that eye color isn't a choice. My green eyes are not something for which I can take credit or blame. We do, however, judge people for murder. As we do draw these sorts of distinctions between different types of human outcomes, we can consider whether it seems reasonable to take information about genetics (and information about the role of the early environment) into account when we are drawing them. And in my view, the answer is yes—the extent to which someone's life outcomes can be traced back to their starting point in life calls into question how much they could have acted differently.

In the movie *Jurassic Park*, there's a scene, before the dinosaurs start eating people, when Jeff Goldblum's character is explaining the "unpredictability of complex systems" to Laura Dern's character. He holds the top of her hand out flat ("like a hieroglyphic") and dribbles a drop of water on her hand. Which way is it going to roll off? She makes a prediction, and they watch where the water slides. Then Goldblum asks her to repeat the exercise: "I'm going to do the same thing, start with the same place again. Which way is it going to roll off again?"

The second time around, the water droplet slides off the opposite side of Dern's hand. Why didn't the water slide off her hand in exactly the same way twice? Goldblum uses the variation in outcome to engage in some professionally inappropriate caressing of Dean's hand, and also to posit that "tiny variations . . . never repeat and vastly affect the outcome."

Two identical twins raised together are like two drops of water starting at the same place. They begin as one zygote, and only later divide into what will become two separate people. They have, by virtue of being conceived with the same egg and the same sperm, a nearly identical DNA sequence (although not *entirely* identical, as developmental mutations can affect one twin but not another).[15] They are fetuses in the same womb. They are typically born at the same time, and they are typically reared in the same house by the same parents who have the same flaws, strengths, and idiosyncrasies. They typically go to the same school and live in the same neighborhood.

Yet "identical" twins are not *exactly* identical for many of their life outcomes. Differences between them—one twin fat, the other thin; one twin afflicted with schizophrenia, the other unscathed—can be as mesmerizing as their similarities.

Researchers typically label these differences between identical twins as the "non-shared environment," abbreviated e^2. The lower the correlation between identical twins, the higher the non-shared environment. Possible e^2 values range from 0, meaning that identical twins are always perfectly alike in their outcomes, to 1, meaning that identical twins are no more similar than two people plucked at random from the population. Thus, e^2 represents differences between people that are *not* due to differences between them in their DNA or in the social circumstances into which they were born. e^2 reflects the degree to which two identical drops of water, beginning in the same place, fall off in different directions.

Eric Turkheimer, my former PhD advisor, proposed that this individuality in human outcomes, which remains after one has considered the constraints of genetics and of family upbringing, is a way of "quantifying human agency."[16] His reasoning is this: We consider someone as having choice and control over an outcome if they *could* have done differently. If people who share the same accidents of birth—who have the same genetics (with the aforementioned qualifications) and the same family upbringing—never actually *do* turn out differently, it becomes harder to imagine that they could have done so. Unpredictability, in his view, becomes a sign of freedom:

> The nonshared environment is, in a phrase, free will. Not the sort of metaphysical free will that no one believes in anymore, according to which human souls float free above the mechanistic constraints of the physical world, but an embodied free will . . . that encompasses our ability to respond to complex circumstances in complex and unpredictable ways and in the process build the self.

In Turkheimer's view, the individual phenotypic space that is not determined by either your genotype or the environmental circumstances defines the boundaries in which your free will gets to play. To borrow a phrase from the philosopher Daniel Dennett, e^2 lets you know how much "elbow room" you have to choose who are you going to be.[17] Turkheimer then lists various human outcomes, pointing out that identical twins are only minimally different for outcomes that we think of as involving little choice and moral responsibility. The e^2 coefficient for height is <0.1. That is, given that one is born to a particular family in a particular time and place with a particular genome, there is little "elbow room" for how tall you are going to be. But that's height—what about social and behavioral outcomes, like education?

The Free Will Coefficient in Education

So, if the extent to which identical twins differ from one another is an indication of the extent to which people *potentially* have agency over their outcomes,[18] let us consider what we actually observe about identical-twin differences in social and economic outcomes, and in the psychological phenotypes that are rewarded (in modern industrialized capitalist societies) with social and economic success.

Turkheimer notes that the e^2 coefficient for IQ is only a bit larger than what is observed for height—0.2 in adulthood, 0.25 in childhood, versus 0.1 for height. But even that marginal difference could be due to greater difficulty with measuring IQ reliably. In studies using statistical techniques to correct for measurement error, the e^2 for intelligence test performance is closer to 0.1. In a classic study of twins separated at birth and raised in separate households, the average difference in intelligence test scores between twins was about equal to the average difference in the test scores of a single person who took the test twice.[19]

That's general cognitive ability, but we can also examine more-basic cognitive processes. General executive function, which I told you about in chapter 6, is the ability to direct and allocate attention. Among children and adolescents, e^2 in general executive function is scarcely greater than zero (< 0.05), while e^2 in processing speed is again comparable to height (0.15).[20]

When we consider academic outcomes, e^2 is *sometimes* larger than what is observed for cognitive abilities, but not uniformly. In our sample of twins in Texas, e^2 was around 0.3 for reading and math achievement test scores in childhood.[21] In the UK, however, achievement test scores from childhood to adolescence show e^2 estimates of <0.15. Similarly, scores on the General Certificate of Secondary Education (GCSE), the nationwide standardized test that is the gateway to university acceptance, show an e^2 of 0.13.[22]

For educational attainment, there is considerable variation across countries and cohorts (from 0.11 to 0.41 across all studies)—with an average across all samples of 0.25. In Scandinavian cohorts, e^2 was smaller (0.17)—again, not that much bigger than what we see for adult height.[23] For income, e^2 is around 0.4 for men (averaging 20 years of income),[24] which is comparable to what is seen for something like depression—but still lower than what is observed for personality.

When reviewing twin research from this perspective, with an exclusive focus on e^2, the nature-nurture debate melts away. We are not trying to parse whether it is their shared outcomes in the social lottery or in the natural lottery of genetic inheritance that leads identical twins raised together to have such similar lives. Rather, we can simply appreciate that, after the joint outcomes of both the social and the natural lotteries are taken into account, the remaining unpredictability in psychological traits and socioeconomic positions is small. Once we account for the powerful effects of luck—both environmental and genetic, together—there is remarkably little territory left for "personal responsibility."

After all, to say that someone is responsible for a life outcome is to imply that, in theory, she could have done differently. Generally, "Could this person have done differently?" is an impossible question to answer. You only have one life to live, so the extent to which you could have been different, or behaved differently, *if only you had so chosen,* is intractable empirically. But following the lives of identical twins raised together tells us that, in practice, people who begin life in the same place, with the same parents and the same ZIP code *and the same genes*, rarely end up with different educational outcomes. They have nearly the exact same executive functioning skills and high-stakes university admission test scores, and fairly similar levels of ultimate educational attainment.

Your genotype, like the social class of your family, is an accident of birth over which you had no control. Your genotype, like the social class of your family, is a type of luck in your life. And the literature on identical twins shows us that, together, the natural and social lotteries are powerful predictors of someone's social position in adulthood, particularly their educational attainment.

The Ideology of Luck

Research in social psychology says that your response to the previous paragraph will vary depending on your politics. One study found that, compared to liberals, conservatives disagree more with statements like, "Successful people are likely to have been lucky in their lives" and agree more with statements like, "People do not need luck to do well in their lives."[25] In another experiment by the same researchers, people were asked to read a passage taken from a commencement address at Princeton University by author Michael Lewis.[26] One version of the passage used Lewis's original words about the role of luck in success ("People really don't like to hear success explained away as luck—especially successful people"); another version replaced reference to luck with references to "help from other people." Political conservatives were particularly likely to disagree with the passage if they read the version attributing success to luck. (Both liberal and conservative participants thought the author of the passage was less likable, less wise, and less admirable when they read the version of the passage attributing success to luck than when they read the version attributing success to help from other people.)

More recently, a Gallup poll found that American supporters of President Trump were less likely to agree that the "rich are luckier" than opponents of Trump (27% versus 38%).[27] About the same proportion of Trump supporters (26%) agreed that income differences between the rich and poor are unfair. This is a strikingly low percentage compared to the global average: across 60 countries, most people (69%) say that income differences in their country are unfair.

We can see conservative squeamishness about acknowledging the role of luck elsewhere, too. In an opinion piece in the *Wall Street Journal*, Heather MacDonald—a conservative writer who authored such books as *The Diversity Delusion* and *The War on Cops*—insisted that "a random roll of the dice [wasn't] sufficient to make today's business titans or their predecessors successful." Rather, "behavioral choices shape life trajectories."[28] (She also claimed that "only the most draconian government leveling could erase" the effects of unequally distributed "innate gifts"—an example of the hereditarian pessimism about the possibility of social change that I discussed in chapter 8.)

The same theme emerged when Dan McLaughlin, another conservative commentator, responded to statements made by Elizabeth Warren while she was campaigning for the US Senate in 2011. One of her speeches, about how public investments were necessary for the success of private business, went viral. McLaughlin described why he thought Warren's point was dangerous to the conservative agenda: it might justify more redistribution of wealth. "If you convince people that success has little to do w/work & merit, you justify more burdens on successful," he tweeted.[29]

Recognizing the conservative aversion to attributing outcomes to luck helps us make sense of two pieces of data that might be surprising, if you are used to thinking of genetics research as necessarily affirming a right-wing worldview. First, conservatives are *less* likely than liberals to attribute people's life outcomes to genetics, particularly for moralized outcomes like sexual orientation and drug addiction.[30] And, second, conservatives are *less* likely than liberals to agree that rich people "are born with greater abilities."[31]

McLaughlin's reason for opposing the ideology of luck was his fear that emphasizing the role of luck increases support for redistribution—and his intuition was correct. People are, in fact, more likely to support redistribution when they see inequalities as stemming from lucky factors over which people have no control than when they see inequalities as stemming from choice.

This connection between the source of inequality—choice versus chance—and people's willingness to redistribute money was shown in a series of fascinating experiments by a team of Norwegian economists.[32] Many of their studies involve variations on an economic game with two parts. In the first part, the "production phase," participants in the study "earn" money by performing a task, such as typing words. Several variables are at play. Some of these variables are generally considered to be under someone's control, such as how many minutes a person spent typing. Some variables are obviously not under someone's control, such as the price that the experimenter sets for each word typed. Some variables, such as how many correct words a person can type in a minute, are ambiguous regarding how much control are person has over them. And some variables pit a person's piece of the pie against the overall size of the pie: for example, points are more valuable, in terms of how much money they are worth at the end of the game, if they are distributed more unequally, such that a person could end up with more money by accepting a more unequal distribution.

In the second part of the game, people—either the players themselves or the spectators—are asked to decide whether everyone gets to keep exactly what they earned or whether some money should be redistributed. People have different preferences in these studies. Some people prefer equality of outcome regardless of the source of inequality; they are radical egalitarians. Some people prefer to let each person keep exactly what they earned in the game, regardless of how it came to be; they are libertarians.

But, most commonly, people distinguish fair from unfair inequality on the basis of whether the source of that inequality was chance or choice: If people end the game with less money because they were unlucky in the price the experimenter assigned their work, then people are more likely to redistribute money to counteract that inequality.

People's sensitivity to the role of luck in economic games is mirrored in their responses to surveys about inequalities.[33] In one survey in Norway, nearly half of people (48%) said that inequalities in income that stem from factors outside of a person's control should be eliminated. Americans, on the whole, are more inequality-accepting than are Norwegians, and political conservatives are more inequality-accepting than political liberals. But across the board, people are more willing to redistribute to equalize outcomes due to luck than redistribute inequalities stemming from factors considered under a person's control.

In economic games designed to measure people's distributional preferences, and in surveys about fair and unfair inequalities, the types of luck that produce unfair inequalities are outside events that happen *to* a person and that constrain the person's overall control over their social and economic outcomes. The experimenter set a low price on your work. You were born to a mother who didn't finish high school.

As we have seen, however, inequalities in these same social and economic outcomes *also* stem from another factor outside of a person's control but internal to the self—genetics.

Wanting to Blame, Revisited

In the first chapter of this book, I told you about a social psychology study in which participants were told about a fictional Dr. Karlsson, who either found that genetic influences accounted for a little (4%)

or a lot (26%) of the variation in how people performed on a math test. In the latter condition, people—particularly political liberals—perceived Dr. Karlsson as being less objective and as holding less egalitarian values. That is, when Dr. Karlsson reported stronger genetic influences on math test performance, he was perceived as believing things like, "It's OK if society allows some people more power and success than others," and as *not* believing things like, "Society should strive to level the playing field to make things just."

The results of this study accord with my own personal experience, as someone who writes and gives talks about behavioral genetic research: acknowledging the existence of genetic influences on socially important outcomes, such as math test performance or educational attainment, is widely perceived to countervail egalitarian values. There are, of course, good reasons for this. Historically, genetic ideas were used by ideological extremists to justify profoundly non-egalitarian social policies, such as limiting immigration from certain regions of the world, forcibly sterilizing people, and even detaining and murdering people.

Even as we recognize *why* certain types of genetic studies are associated, both historically and in popular imagination, with extremist views about human superiority and inferiority, the research that I've described in this chapter points to a far different framing. A person's genotype is *randomly* chosen from the possible genotypes they could have inherited from their parents—it is a matter of luck. On average, it is people who are politically conservative who are reluctant to say that luck plays a role in people's success. Similarly, people are more likely to reject information about genetic influences when they want to blame or otherwise hold people responsible for their (mis)behavior. But when inequality *is* seen as stemming from lucky factors over which people have no control, both conservatives and liberals are more likely to see those inequalities as unfair and to support redistribution to equalize resources.

Considered together, these points are the ingredients of a new synthesis: Genetics is a matter of luck in people's lives. Appreciating the role of genetic luck in people's educational and financial success undercuts the blame that is heaped on people for not "achieving" enough and might, in fact, bolster the case for redistributing resources to achieve greater equality.

On the flip side, rejecting information about genetic influences on social and economic outcomes might have the unintended side effect of *increasing* the blame heaped on anyone who failed to advance in their education and who is doing poorly in an economy that benefits only "skilled" workers. Stigmatization of the poor as blameworthy was the concern of Michael Young, the British socialist who first coined the term "meritocracy" to describe a dystopian future. Over forty years after first using the word "meritocracy," Young reflected ruefully on how those who have not done well in school are held responsible for their own lack of success: "It is hard indeed in a society that makes so much of merit to be judged as having none. No underclass has ever been left as morally naked as that."[34]

11

Difference without Hierarchy

My older child struggled to talk. When he was two, he had only a few words, and the pediatrician offered us reassurances—he was fine, just be patient, boys can be late talkers. Six years later, he's been in hours and hours of speech therapy every week. Therapists have reached into his mouth to hold down the front of his tongue so he can say "cookie" and "go." He practices holding his jaw and rounding his lips and saying the correct number of syllables before he draws another breath.

During his speech therapy appointments, I sit in the waiting room and read with my younger child, who was a precocious talker. Her speech development felt miraculous in comparison with her older brother's. What had to be relentlessly practiced with one child emerged with seeming effortlessness in another.

Why can one of my children talk with ease, while the other one labors to be understood? No one can give me a definitive answer. But I can look to twin studies and see that speech problems are over 90 percent heritable. Most of the differences between children in their ability to articulate words are due to genetic differences between them. The genetic influences on speech problems also appear to influence motor skills more generally, a scientific finding that comports with my personal experience. I watched my late talker struggle to learn to crawl, to walk, to ride a scooter.

Of course, the high heritability of speech impairment does not obviate the importance of the environment. The only available interventions for speech problems are environmental ones; no one is CRISPR-ing the genome of the late-talking three-year-old. And we can, unfortunately, find numerous examples of abused, neglected, or abandoned children who were deprived of verbal interactions in their early lives, with devastating results. But against the backdrop of the normal linguistic environment provided in my home, it is likely that my children differ in their verbal development because they differ in which genes they happened to inherit from me and their father.

Discussing the heritability of speech impairment is not, in my personal experience, controversial. Most speech therapists will inquire about one's family history of speech problems. Most parents of more than one child can observe how differently their children's speech and language development unfolds. Looking at how my children differ in their ability to articulate words, I can easily see the capricious hand of nature. When it comes to inheriting whatever combination of genetic variants allows one to pronounce a word like "squirrel" by the age of three, my daughter was lucky. My son was not.

Given that the combination of genetic and environmental factors that resulted in her typically developing speech and language abilities were entirely out of her control, it would be very strange to say that my daughter did anything to *earn* her verbal precocity. Her speaking in complex sentences at any early age doesn't make her *good*. If anything, the praiseworthiness belongs to my son, who brings the same deliberate, effortful attention to breath support and intonation to his daily conversations as an opera singer brings to a performance at the Met.

In the late afternoon, on our way home from speech therapy, we stop at a stoplight near a freeway underpass in south Austin. The underpass hosts a growing homeless encampment—sleeping bags, tents, wheelchairs, shopping carts piled high with tattered belongings. In the winter, the population swells. In the summer, the few remaining camp inhabitants swelter as temperatures exceed 100 degrees. I keep bottles of water in the car for the men (and it's almost always men) who are holding cardboard signs at the intersection. I pass water through my car window to men with sunbaked hands, and my children ask questions.

"Will ghosts get them at night?"

"Why don't they have houses?"

"Why do *we* have a house?"

One of the awesome responsibilities of parenthood is that I can tell them whatever I want. I could say that those men made bad choices. I could quote the Bible verse from Thessalonians that was quoted to me as a child: "The one who is unwilling to work shall not eat." But no matter how many times we have this conversation, I always end up saying a version of the same thing: That we are lucky. That Mama is lucky to have a job that pays her money, and that's how we buy clothes and food and toys and our house. That some people were unlucky in their life. That being unlucky shouldn't mean that you sleep under a bridge, but we adults don't always share enough of our money so that everyone has a house.

(My daughter asks another question: "Why do people have selfish in their heart?")

Just as genetic differences between people create differences between them in their likelihood of developing speech problems, so, too, do genetic differences between people create differences between them in their likelihood of being homeless. There has not been a GWAS or twin study of homelessness, but the statement is almost certainly true. About 20 percent of the homeless population has a serious mental illness, like bipolar disorder or schizophrenia.[1] About 16 percent are estimated to have a serious substance use disorder, such as alcoholism or opioid addiction. Ultimately, the cause of homelessness is *not being able to afford housing.* And if people had not inherited certain genetic variants, then the probability of them experiencing all these things—mental illness and addiction and poverty—would be different. Those of us who have not experienced the challenges of psychosis or addiction or deep poverty *are* lucky. Some of that luck is circumstantial; some of that luck is embodied.

Two Concerns about Genetic Research

Genetic differences between people create differences between them in their likelihood of having speech and language problems. Genetic differences between people create differences between them in their likelihood of being homeless. The first sentence is not particularly controversial; the second one almost definitely is.

But why?

The bioethics scholar Erik Parens summarized what I believe are the two core concerns that stoke controversy, even outrage, about connecting genetic differences between people to social inequalities like poverty and homelessness: "By investigating the causes of human differences, people worry, *behavioral genetics will undermine our concept of moral equality.* . . . Unfortunately, there is an old and perhaps permanent danger that inquiries into the genetic differences among us will be *appropriated to justify inequalities in the distribution of social power*" (emphases mine).[2]

Parens's summary of why people worry about (some) behavioral genetic findings bears striking similarities to Elizabeth Anderson's definition of inegalitarianism, which I first mentioned in the introduction.[3] She writes: "Inegalitarianism asserted the justice or necessity of basing social order on a hierarchy of human beings, ranked according to intrinsic worth. *Inequality referred not so much to distributions of goods as to relations between superior and inferior persons.* . . . Such unequal social relations *generate, and were thought to justify, inequalities in the distribution of freedoms, resources, and welfare.* This is the core of inegalitarian ideologies of racism, sexism, nationalism, caste, class, and eugenics" (emphases added).

Here, again, we see the same two core concerns. First, to link biological difference to social inequalities is to allege that some people are superior to inferior others—a hierarchical view of human worth that starkly contrasts with the egalitarian idea of human moral equality. And, second, such a hierarchical view of humanity will justify inequalities. Rather than poverty and oppression being problems to be solved, these inequalities will be seen as right and natural consequences of human biological superiority.

These two concerns might seem inescapable. When I say that people differ genetically, and that these genetic differences have consequences for their education and social class, for their income and employment and chances of ending up homeless, it might feel impossible for that statement to be interpreted in any way other than as an assertion about a hierarchy of human worth and the inevitability—rightness, even—of poverty.

As the poet and activist Audre Lorde explained, "Much of Western European history conditions us to see human differences in simplistic opposition to each other: dominant/subordinate, good/bad, up/down, superior/inferior." As a result, she argues, "too often, we pour the energy needed for recognizing and exploring difference into pretending those differences are insurmountable barriers, or that they do not exist at all."[4] The eugenicist ideology is to claim that genetic differences are insurmountable barriers to equality; too often, the response to eugenicist ideology is to pretend that genetic differences do not exist at all.

But we don't *always* talk about genetic differences between people in terms of hierarchies of inferior and superior people—and these examples can be instructive. When I say that my children differ genetically and that these genetic differences have real consequences for their ability to talk, I certainly am not implying that one of my children is "superior" or "inferior" to the other one. Verbal ability is *valued*, but having strong verbal ability doesn't make one of my children more *valuable*. The genetic differences between them are meaningful for their lives, but those differences do not create a hierarchy of intrinsic worth.

And those differences don't justify my entrenching inequalities by investing different levels of resources in each of their lives. If anything, the opposite is true. In light of the differences between them, treating my children exactly the same in the interest of "fairness" would feel absurd. Instead, I invest hours and hours more per week in the speech and language development of the child who struggles, because that additional training and investment is what he needs.

In the context of a specific (dis)ability like childhood speech production, and in the context of a within-family sibling comparison, how we make sense of genetic differences between people slips free from the noose of inegalitarianism. We can, perhaps, talk about genetic differences between people *without* slotting them into a hierarchy of human worth. We can, perhaps, acknowledge that people are not born with an equal statistical likelihood of experiencing certain life outcomes, *without* justifying the differences in life outcomes among them as inevitable and natural.

Why is this difficult to do (and it *is* difficult) when we are considering genetic differences in relation to social inequalities? In contrast to speech impairments in childhood, a concept like "intelligence" is more easily seen as *inherently* hierarchical. When we are talking about DNA, the word "worth" has a subtly dangerous double meaning: someone's net worth, in terms of the market value of their financial assets, can be too easily conflated with their intrinsic worth as humans. When net worth is associated with genetic differences, it is tempting to slip into thinking that intrinsic worth is also tied to one's DNA. In this chapter, I want to consider the historical reasons behind why genetic research on some human outcomes—scores on intelligence tests foremost among them—automatically activate notions about human inferiority and superiority. I then want to consider alternative examples of human phenotypes—such as height, deafness, and autism—where genetic research has been largely embraced rather than rejected as dangerous. Can we look to these examples to broaden our intuitions about whether genetic research on social inequality is *necessarily* dangerous?

Socially *Valued*, Not Inherently *Valuable*

The tendency to see intelligence (as measured on standardized IQ tests) and educational success, perhaps more than any other human phenotypes, in terms of a hierarchy of inferior and superior persons is not an accident. It is an idea that was deliberately crafted and disseminated. As the historian Daniel Kevles summarized, "Eugenicists [in the early twentieth century] identified human worth with the qualities they presumed themselves to possess—the sort that facilitated passage through schools, universities, and professional training."[5] And this equation is most clearly on display in the history of intelligence testing.

The first intelligence tests were created by a pair of psychologists, Alfred Binet and Theodore Simon, who had been tasked by the French government with developing a means to identify children who were struggling in school and needed additional assistance. The resulting Binet-Simon scale asked children to do a series of practical and academic tasks that were typical of everyday life. An eight-year-old child was asked, for instance, to count money, name four colors, count backward, and write down dictated text.

The key advances of the Binet-Simon scale didn't lie in which specific tasks they asked of children, but rather in two innovations. First, the *same* tasks were asked of everyone (*standardization*). Second, the same tasks were administered to a large number of children, permitting statements to be made about how the *average* child of a certain age performed and how the performance of any one child compared to that age-graded average (*norming*).

Any parent who has ever consulted a growth chart to see whether their child is gaining enough weight, or who has ever asked a teacher whether their child's reading is keeping up with the rest of the class, will immediately grasp the power of norming. You can look around at your friends' children, or try to recall what your older children were like at that age, but you don't really know—what *is* the typical weight for an 18-month-old? How many words can the average 6-year-old sight-read? A properly constructed set of norms won't tell you *why* a child isn't gaining weight or is struggling to read. Norms for one set of tasks won't tell you whether there are other socially valued skills that *aren't* being measured. But norms *will* give you some comparative data that is grounded in something other than people's subjective intuitions about what children can and can't do.

Tragically, the Binet-Simon scale was nearly immediately appropriated as a quantitative metric that justified the inegalitarianism that already characterized American society. Psychologists *discovered* some things about measurement: if you ask children to perform a finite number of tasks, older children can do more things than younger children; children differ in the rate at which their performance on those tasks improves; and differences in performance on a small number of tasks can be informative about which children will struggle with a much broader set of learning tasks they face in their lives. And then psychologists *invented* another idea: that performance on those tasks could be used to tell you which people were better than other people.

In 1908, the American psychologist Henry Goddard imported the Binet-Simon tests from France to the United States, translating them to English and using them to test thousands of children. Goddard published the results in a 1914 book, *Feeble-Mindedness: Its Causes and Consequences*.[6] In it, Goddard alleged that the so-called "feebleminded" were physically distinct: "There is an incoordination of their movements and a certain coarseness of features which do not make them attractive, but in many ways suggest the savage."

More damningly, people with low scores on the early intelligence tests were alleged to be deficient *morally*. According to Goddard, they lacked "one or the other of the factors essential to a moral life—an understanding of right and wrong, and the power of control." At the same time, "the folly, the crudity" of immoral behavior, including all forms of "intemperance and . . . social evil," were considered "indication[s] of an intellectual trait." Combining intellectual, physical, and moral deficits, the overall picture that Goddard painted of "feeblemindedness" was appalling in its dehumanization: the "feebleminded" man or woman was "a more primitive form of humanity," a "crude, coarse form of the human organism," "a vigorous animal."

In this way, Goddard and his contemporaries positioned intelligence test scores as a numerical referendum on one's human value. People with low scores were "primitive" humans, animal-like in their physical savagery and lack of moral responsibility. As the historian

Nathanial Comfort summarized, "IQ became a measure not of what you do, but of who you are—a score for one's inherent worth as a person."[7] It was *this* concept—not "How many questions do you get right on a standardized intelligence test?" but "How primitive is your humanity?"—that was then attached to ideas about heredity and genetic difference.

As a clinical psychologist who has overseen the administration of literally thousands of IQ tests, I found reading Goddard's book a deeply uncomfortable experience. Goddard was one of the founders of American psychology, a group that transformed the field into an experimental science rather than a subfield of philosophy. He helped to draft the first law mandating the availability of special education services in public schools. *Atkins v. Virginia*, the 2002 Supreme Court decision that found that people with intellectual disabilities should not be subject to the death penalty, would have been cheered by Goddard, who was the first person to give legal testimony that people with low intelligence had reduced criminal culpability. Anyone working as a forensic or clinical or school psychologist today is working in a field that Goddard helped to create (just as anyone doing *any* statistical analysis is inescapably indebted to Galton, Pearson, and Fisher). Yet Goddard worked deliberately to establish what I consider an abhorrent idea—that intelligence test scores are a measure of someone's worth.

Fast-forward a century, and the idea that intelligence test scores could be used as a referendum on someone's very humanity continues to haunt any conversation about them. In 2014, for instance, the writer Ta-Nehisi Coates, angry that people were "debating" the existence of genetically caused racial differences, made it clear that he considered questions about one's intelligence to be inseparable from questions about one's humanity: "Life is short. And there are more pressing—and actually interesting—questions than 'Are you less human than me?'" Other writers responded with apparent bewilderment at Coates's statement (e.g., "It genuinely grieves me," wrote Andrew Sullivan).[8] But such bewilderment belies willful ignorance about the history of intelligence testing. Coates's rhetorical question—"Are you less human than me?"—was the exact question that the early proponents of intelligence testing were asking in earnest.

No discussion of intelligence and educational success can ignore this history. In fact, given this history, multiple scholars have advocated abandoning standardized testing and the concept of "intelligence" entirely. In this view, there is no legitimate way to study intelligence, even within a racial group, because the concept of intelligence is itself an inherently racist and eugenic idea. The historian Ibram X. Kendi, in his book *How to Be an Antiracist*, gave a trenchant expression of this concern: "The use of standardized tests to measure aptitude and intelligence is one of the most effective racist policies ever devised to degrade Black minds and legally exclude Black bodies."[9]

Thus, even if molecular genetic studies of intelligence and educational attainment are focusing their attention exclusively on understanding differences between *individuals* within European ancestry populations, some consider the work to *still* be the fruit of the poisoned tree.

But other writers paying attention to race and racism have concluded that, despite their original intents, IQ tests are nonetheless valuable tools for understanding the effects of discriminatory policies. As Kendi himself describes, identifying racial inequity is critical to fighting what he calls "metastatic racism":

> If we cannot identify racial inequity, then we will not be able to identify racist policies. If we cannot identify racist policies, then we cannot challenge racist policies. If we cannot challenge racist policies, then racist power's final solution will be achieved: a world of inequity none of us can see, let alone resist.

The importance of documenting racial inequities in health outcomes like life span, obesity, and maternal mortality is obvious: How are we to close these disparities, to investigate how policies affect them, if we cannot measure them? For instance, knowing that desegregating Southern hospitals closed the Black-White gap in infant mortality and saved the lives of thousands of Black infants in the decade from 1965 to 1975[10] requires, at a minimum, being able to *quantify* infant mortality.

Documenting racial inequities in health means documenting racial inequities in every bodily system—including the brain. And some racist policies harm the health of children by depriving them of the social and physical environmental inputs necessary for optimal brain development, or by exposing them to neurobiological toxins.

Consider lead. In 2014, when the city of Flint, Michigan, switched the source of its drinking water supply from Lake Huron to the Flint River, Flint residents—the majority of whom are Black—immediately complained about the switch: an early story by CBS News was titled, "I don't even let my dogs drink this water."[11] The new water supply was corrosive. As it flowed through the antiquated lead pipes of the city's water system, lead leached into the drinking water. In areas of the city with particularly high lead levels, the percentage of children with elevated blood lead levels nearly tripled, to over 10 percent.[12] Those areas with the highest exposure to lead were also the areas with the highest concentration of Black children. The confluence of factors visiting harm on these children led the Michigan Civil Rights Commission to conclude that the lead poisoning crisis was rooted in "systemic racism."[13]

What tool is used to measure the neurotoxic effects of lead? IQ tests. The IQ deficits that result from lead exposure prevent researchers and policymakers from shrugging off the effects of lead as temporary or trivial. And that is just one example. In her book, *A Terrible Thing to Waste*, Harriet Washington documents how people of color are overwhelmingly more likely to be exposed to environmental hazards like toxic waste and air pollution. Moreover, she argues that IQ tests, by providing a numerical metric for a child's ability to reason abstractly, are currently an irreplaceable tool for quantifying the perniciousness of what she terms "environmental racism:"[14] "In today's technological society, the species of intelligence measured by IQ [tests] is what's deemed most germane to success. . . . IQ is too important to ignore or wish away."[15]

Washington is right: the skills measured by IQ tests, while certainly only representing a fraction of possible human skills and talents, cannot be wished away as unimportant. In Western high-income countries like the United States and the UK, scores on standardized cognitive tests (including scores on the classic IQ tests, and also scores on tests used for educational selection, like the SAT or ACT, which are highly correlated with IQ test scores[16]) statistically predict things that we care about—including life itself. Children who scored higher on an IQ test at age 11 are more likely to be alive at age 76—and, no, that relationship cannot be explained by the social class of the child's family.[17] Students with higher SAT scores, which are correlated as highly as 0.8 with IQ,[18] earn higher grades in college (especially after one corrects for that fact that good students select more-difficult majors).[19] Precocious students with exceptionally high SAT scores at a young age are also more likely to earn a doctorate in a STEM field, to hold a patent, to earn tenure at a top-50 US university, and to earn a high income.[20]

Washington's quest to reclaim intelligence tests as a tool to combat environmental racism mirrors the efforts of other scholars of color and feminist scholars who have argued that quantitative research tools can be used to challenge multiple forms of injustice. The feminist Ann Oakley, for example, argued that "the feminist case" for abandoning quantitative methods was "ultimately unhelpful to the goal of an emancipatory social science."[21] Similarly, Kevin Cokley and Germine Awad, my colleagues at the University of Texas, affirmed that "some of the ugliest moments in the history of psychology were the result of researchers using quantitative measures to legitimize and codify the prejudices of the day."[22] They went on to argue, however, that, "quantitative methods are not inherently oppressive," and can, in fact, "be liberating if used by multiculturally competent researchers and scholar-activists committed to social justice."

Intelligence tests were positioned by eugenicists as a measure of someone's inherent worth, with the resulting hierarchy of inferior and superior humanity conveniently ratifying the ugliest suppositions of a racist and classist society. Intelligence tests measure individual differences in cognitive functions that are broadly relevant, in our current societies, to people's performance at school and on the job, even to how long they live. The challenge is to reject the former without denying the latter. Like a measure of a child's speech impairments, intelligence tests don't tell you that a person is *valuable*, but they do tell you about whether a person can do (some) things that are *valued*.

Good Genes, Bad Genes, Tall Genes, Deaf Genes

The distinction between *inherently valuable* and *socially valued* might be unfamiliar as applied to our understanding of intelligence test scores, but we can look to three other phenotypes where it is more typical: height, deafness, and autism spectrum disorders.

In chapter 2, I told you about the towering NBA player Shawn Bradley, who inherited an extraordinarily large number of height-increasing genetic variants. Height is perhaps the simplest example of how genetic differences between people can be filtered through a particular cultural and economic system, resulting in differences in socioeconomic status. Over the course of his NBA career—a career that would have been impossible if he had been 5′11″ rather than 7′6″—Bradley earned nearly $70 million. (It's not just basketball players who benefit economically from height-increasing genetics. One analysis found that, in the general population, each inch of height was associated with about $800 greater annual earnings.)

When Bradley says he feels "lucky" that he inherited height-increasing genetic variants, he clearly means he was the beneficiary of *good* luck. Indeed, in ordinary English, "lucky" implies a value judgment—good luck, not bad; feast, not famine. But "good" and "bad" are value judgments that we cannot always apply to DNA. If you get one copy of a particular version of the *HBB* gene, then your body is more resistant to malaria. If you get two copies, however, you will develop sickle-cell anemia, periodically depriving your body of oxygen and eventually killing you. There is no clear answer to whether inheriting a mutated version of *HBB* is good luck or bad luck.

Perhaps no community has challenged notions about what constitutes "good" genetic luck more than the Deaf community. The capital "D" in Deaf is used to represent Deafness as a distinct subculture that shares a common language (American Sign Language), in contrast to lowercase-"d" deafness, which is a condition defined by the inability to hear. As Carol Padden and Tom Humphries wrote in

Deaf in America: Notes from a Culture, Deaf culture is "not simply a camaraderie with others who have a similar physical condition, but is, like many other cultures in the traditional sense of the term, historically created and actively transmitted across generations."[23] You diagnose deafness with an audiological exam; you assess whether or not someone identifies as Deaf the same way you assess whether or not someone identifies as Dutch.

About 1 in 1000 infants is born deaf. Events like being deprived of oxygen or being infected with cytomegalovirus or rubella can cause hearing impairment at birth, but about half of cases of congenital deafness are genetically caused.[24] The genetic architecture of congenital deafness is much simpler than that of something like height: Most cases are monogenic rather than polygenic, meaning that they are caused by a mutation in a single gene.

In the United States, the most common genetic cause of congenital deafness is a recessive variant of *GJB2*. This gene codes for something called connexin 26, which allows for small molecules like potassium to be channeled between adjoining cells.[25] "Recessive" means that the variant typically recedes into the background; you can carry the variant and never know it. But if you inherit two copies of the recessive variant, one from each parent, then its effects no longer recede, and the infant is born deaf. Because the variant is recessive, the intergenerational transmission of deafness works like Mendel's pea plants—the child of a hearing person and a deaf person will likely be hearing, except in the rare case that the hearing person also carries the recessive allele, in which case there is *still* a 50/50 chance the child will be hearing. The child of two deaf parents might also be hearing if each parent's deafness is caused by a different genetic mutation.

Given that the most likely outcome of the genetic lottery is a hearing child, some Deaf parents stack the odds in favor of their desired outcome—a deaf one. In the early 2000s, Candace McCullough and Sharon Duchesneau, who were both born deaf, made international headlines for choosing a sperm donor—a friend whose family has been deaf for five generations—with the intention of conceiving a deaf child. As hoped, both of their children were indeed born deaf.

The *Journal of Medical Ethics* sought to explain their rationale: "Like many others in the deaf community, the couple don't view deafness as a disability. They see deafness as a cultural identity and the sophisticated sign language that enables them to communicate fully with other signers as the defining and unifying feature of their culture."[26]

As technology for measuring the genome improves by leaps and bounds, a new way to stack the genetic deck has become possible—pre-implantation genetic diagnosis, or PGD. PGD allows couples who have used IVF to create several embryos to screen those embryos genetically, in order to select which ones to implant and which ones to discard. PGD is most commonly discussed as a potential means to create so-called "designer babies," i.e., embryos that have been selected for characteristics like height or eye color, which might be considered socially desirable but are not medically necessary. But it also raises the possibility of "negative" selection— selection for characteristics that most prospective parents would find undesirable, like deafness. A landmark survey of fertility clinics in the US found that a small number of clinics (3%) admitted to using PGD to help parents select embryos *for* a disease or disability.[27] Similarly, a survey of Deaf parents found that a small minority would consider terminating a pregnancy if a genetic test found that the fetus would be hearing.

Negative selection using PGD is illegal in the UK, where the Human Fertilisation and Embryology Act prohibits the use of PGD to select any embryo with a genetic abnormality that would cause "serious physical or mental disability." The Act was protested by some in the Deaf community who took offense at being labeled "abnormal," objected to the idea that their condition was so serious that it would be better if they had never been born, and considered it an infringement on their freedom to determine what type of family they wanted.[28] Paula Garfield and Tomato Lichy, a British couple who had one deaf child and were considering using IVF to conceive another, told *The Guardian* newspaper: "Being deaf is not about being disabled, or medically incomplete—it's about being part of a linguistic minority. We're proud, not of the medical aspect of deafness, but of the language we use and the community we live in."[29]

The legal and ethical issues raised by the question of whether Deaf parents should be permitted to pursue the birth of a deaf child—through selection of a sperm or egg donor, through selective abortion, or through pre-implantation genetic testing—are myriad and thorny, and I will not attempt to resolve them here. My goal is simpler: I want to point out how the Deaf community, in advocating for their right to use reproductive technology to conceive a deaf child, nudges us to imagine a different sort of rhetorical relationship between genetic difference, the role of luck in human affairs, and social (in)equality.

As I described above, biological theories of human difference, particularly genetic theories, are widely perceived to be dangerous. If I *say* that people differ genetically in ways that influence their intelligence or social position, then it's easy to *hear* a different claim—that existing social inequalities are natural, inevitable, unfixable, and just. If, then, you look around and see social inequalities that seem very unjust, if you can envision a different world in which they are rectified, then the temptation is to push back on the biological claim—either the research is wrong or the research simply should not be done.

But we don't do this with deafness. That is, we don't reflexively cringe at simply stating that there are biological differences between people that cause differences in their ability to hear.

It is not as if deafness does not have its own history of eugenic atrocity. In Nazi Germany, around 17,000 deaf adults were sterilized; around 2,000 deaf children were murdered; forced abortions were performed on women suspected of carrying a deaf fetus.[30] Even now, many members of the Deaf community see using reproductive technology to select against deaf sperm donors, and against embryos and fetuses who will be born deaf, as a form of genocide.[31]

Yet despite this eugenic history, no one denies that genes can cause deafness. Among other things, the simpler genetic architecture of deafness has ensured that there is simply no rational debate about that question.

Deafness is also relevant to social inequality and social position. Deafness produces social disadvantage; to choose an embryo with deaf genes is to choose a child who is more likely to struggle, academically and economically and occupationally, relative to a hearing child. We are comfortable, on the whole, saying that genes cause deafness, and that deafness makes it harder for a child to succeed academically when in a school system where spoken language predominates, and that academic obstacles in childhood can result in less economic and professional success in adulthood.

Genetic influences on deafness work differently, of course, than genetic influences on intelligence or educational attainment or income. The genetic architectures are different; the mechanisms that connect genes to the outcome are different. But there are also stark differences in the *interpretive frameworks* surrounding these empirical questions. In the Deaf community's quest to be appreciated as different, rather than denigrated as defective, recognition of genetic difference in socially valued traits can exist side by side with an egalitarian insistence that all men are created equal. Such harmonious coexistence might seem difficult to imagine when we are talking about genetic differences relevant to educational outcomes, but there are three ideas we can borrow from the discourse on being deaf/Deaf.

First, the genes that cause deafness are seen—appropriately so—as morally arbitrary. There is nothing morally praiseworthy or blameworthy about being born with one variant of *GJB2* or another. People don't deserve rewards or punishment because they inherited one or two copies of an autosomal recessive allele. In fact, the eugenic tendency of projecting "good" and "bad" value labels down into the genome is steadily and persistently undermined by the insistence of a minority that would perhaps select *for* the very same genetic variants that the majority finds less preferable.

Second, deafness *itself* is seen as morally arbitrary. The Deaf are not more or less virtuous than the hearing, or vice versa. The ability to transduce vibrations in the air into electrical signals that are sent to the temporal lobes of one's brain—that is a basic psychophysiological process, over which people have little to no conscious control or responsibility. Hearing is a functioning, not a virtue.

Third, it is on the basis of their differences in functioning—differences, again, that are caused by genes (*or* environments, like hypoxia at birth) over which an individual has no control—that the Deaf community makes claims on the rest of society. And what the Deaf community is claiming is not compensatory pity for their misfortune. As Elizabeth Alexander put it, "the Deaf . . . want to make claims on the hearing in a manner *that expresses the dignity they see in their lives and community*, rather than in a manner that appeals to pity for their condition" (emphasis added).[32]

On the whole, the relationship between the genetics of deafness and the politics of Deafness is consistent with the ideas of the political philosopher John Rawls, who argued:[33]

> The natural distribution is neither just nor unjust. . . . These are simply natural facts. What is just and unjust is the way that institutions deal with these facts. Aristocratic and caste societies are unjust because they make these contingencies the ascriptive basis for belonging to more or less enclosed and privileged social classes. The basic structure of these societies incorporates the arbitrariness found in nature. But there is no necessity for men to resign themselves to these contingencies. The social system is not an unchangeable order beyond human control but a pattern of human action.

We can connect every sentence in this passage to an aspect of deafness / Deafness. The genetic lottery that produces deaf children or hearing children is a "natural fact" that we can no more criticize as fair or unfair, as just or unjust, than we can be morally outraged that lightning struck in our backyard and not our neighbor's. This "arbitrariness found in nature," however, need not—indeed, should not—be a contingency to which we are resigned. Deafness need not produce an unchangeable order, in which the Deaf are ascribed to a permanent underclass, to a life of poverty and social exclusion. Rather, through laws like the Americans with Disability Act, we have changed the patterns of human action, so that those who are on one end of the natural distribution of hearing ability can more fully participate, as equals, in economic and social life.

We see this Rawlsian idea similarly at work in the "neurodiversity" movement, which centers on people with autism spectrum disorders (ASDs). "On the spectrum" has become part of the modern lexicon, a schoolyard taunt and an armchair explanation for odd behavior. The phrase has become so familiar that the metaphor underlying the word "spectrum" is often forgotten. But *spectra* are literally rainbows, the separation of light into components with different wavelengths that are perceived by the human eye as different colors, which shade one into the next. The spectrum metaphor captures the simultaneity of continuous variation and the human need to clump our experiences into categories.

The existence of an autism spectrum means that many autistic-spectrum adults do *not* suffer from the impairments of functioning characteristic of severe autism (harming oneself, being unable to use the toilet independently, being entirely nonverbal). But they might still identify as part of the broader autism community. In the past decade, high-functioning, on-the-spectrum people, and their relatives and supporters, have reshaped public discourse around autism under the banner of neurodiversity. Among other goals, neurodiversity advocates argue that the cognitive and behavioral features of autism spectrum disorders (and other syndromes like ADHD) are not necessarily bugs, but are rather potential features, of the human cognitive machinery. The neurodiverse might, in the right context, have potentially rare and valuable skills. But, even if they don't have savant abilities in any area, they also make claims on society to express, to return to Elizabeth Anderson's words, "the dignity they see in their lives and community."

In fact, there are an increasing number of examples where the "pattern of human action," as Rawls put it, has been changed to include people with ASDs more fully in occupational and economic life. Some militaries, for example, provide intensive training to teenagers with ASDs, so that young people who have heightened attention to visual detail and pattern can be put to use scanning satellite images.[34] An article in the *Harvard Business Review* proclaimed that "neurodiversity is a competitive advantage," and advised tech companies to "adjust their recruitment, selection, and career development policies to reflect a broader definition of talent."[35] Following this advice, Auticon, a technology consulting business that conducts quality checks for websites and software, specializes in employing people with autism spectrum disorders, who then define the workplace culture.[36] As a result, success in the office depends much less on the ability to read tacit social cues, and management pays more attention to the intensity and consistency of physical stimuli, such as fluorescent lighting and paint color.

As with the Deaf community, the neurodiversity movement makes no attempt to minimize the influence of genetics on autistic spectrum disorders. In fact, genetic influences on autism are taken as foundational. For instance, a blog in *Psychology Today* by someone who identifies as having Asperger's *defined* "neurodiversity" as "the idea that neurological differences like autism and ADHD are the result of normal, natural variation in the human genome."[37] This embrace of genetic research extends to the general public: by now, few people other than anti-vaccination extremists would argue that ASDs are not influenced by genes.

Recognizing that genetics are important for understanding who is tall, or who develops autism, or who is born deaf, is largely uncontroversial. These communities don't stake their claims to equity and inclusion on genetic sameness. Genes are not always a problem to be fixed, or the only problem to be fixed. *People* are not the problem to be fixed. The problem to be fixed is society's recalcitrant unwillingness to arrange itself in a way that allows them to participate.

In the same vein, we can recognize that genetics are important for understanding who develops the cognitive abilities and non-cognitive skills that are valued in the formal education systems of high-income countries. People need not stake their claims to equity and inclusion on genetic sameness, or on the irrelevance of genetic difference for human psychology. Rather, the problem to be fixed is society's recalcitrant unwillingness to arrange itself in ways that allow *everyone*, regardless of which genetic variants they inherit, to participate fully in the social and economic life of this country.

The question becomes, then: How can public spaces and working conditions and access to medical care and legal codes and social norms be reimagined, such that the "arbitrariness of nature" is not crystallized into an inflexible caste system? This is the critical overarching question I want to bring to our discussion of policies in the post-genomic age, which we will consider in the final chapter.

12

Anti-Eugenic Science and Policy

Parasite, a South Korean film by the director Bong Joon-ho that won the Oscar for Best Picture in 2020, is not a movie for the faint of heart. In one scene, a man hiding from debt collectors is revealed to have secretly lived in a windowless basement bunker for years. In another, a torrential rainstorm floods a poor family's semi-basement apartment, filling their home chest-high with brown sewage water. Left without any recourse, the daughter in the family sits on top of the overflowing toilet in the family's one shared bathroom and lights up a cigarette.

These characters and their desperate circumstances are united by their relationship to the wealthy Park family. The Park family matriarch props up her bare feet in the back of a car chauffeured by a man whose home was just destroyed by flooding and comments gaily about how the rain cleared up the pollution, about how the day is perfect for an impromptu party. The character on screen wrinkles her nose in disgust at the scent of her chauffeur, who spent the night in a shelter for displaced people. The audience wrinkles their noses at her oblivious callousness.

Alternating between comic and grotesque, *Parasite* cast an unflinching spotlight on class inequality—the sort of class inequality that, critics fear, will be naturalized and entrenched by genetic research on social and behavioral outcomes. The patriarch of the Park family, after all, is a clean-cut exemplar of the meritocratic ideal, working long hours at a technology company before coming home to his wife and two children. He and his wife talk about the people who chauffeur the cars of the rich but take the subway home as repugnant others who smell like old turnips. How convenient it would be for the Park family, who are obscenely unaware both of their own expansive privilege and of the daily humiliations suffered by their employees, to be told by "science" that their servants are "naturally" inferior.

This is the specter of eugenics: that genetics will be used to establish a "hierarchy of human beings, ranked according to intrinsic worth" and that this hierarchy will be used to produce "inequalities in the distribution of freedoms, resources, and welfare." [1] The former is the core of eugenic ideology. The latter is the effect of eugenic policy.

For decades, scientists such as myself, who both study genetic influences on social behavior and who have egalitarian values, have attempted to fight the specter of eugenics by making arguments about what we should *not* do. Indeed, much of this book has been taken up with these arguments. We should *not* interpret genetic influences as deterministic. We should *not* give up on the possibility of social policy to bring about social change. We should *not* confuse an outcome being socially *valued* with a person being *valuable*. But if we are not using genetic research to feed eugenic ideology and eugenic policy, what *should* we do with it?

One approach is to sweep genetic research under the rug, ignoring a large and remarkably consistent body of scientific knowledge, lest the eugenics genie be let out of the bottle. This is a mistake, analogous to the mistaken ideology of colorblindness. Claiming to "not see race" doesn't make the power of race and racism go away. Rather, failing to recognize a systemic force that creates inequalities permits them to continue under a veil of neutral passivity. Creating a just social order requires antiracism, not colorblindness. [2] Similarly, claiming that genetic differences between people are meaningless does not make the power of the genome go away. Rather, failing to recognize the genetic lottery as a systemic force that creates inequalities does exactly what eugenic ideology would want—permits those genetically associated inequalities to persist as "natural" rather than being critically examined. Creating a just social order requires *anti-eugenics*, not gene-blindness. We must take up the question raised by the sociologist Ruha Benjamin: "How might technoscience be appropriated and reimagined for more liberatory ends?"[3]

Eugenic ideology has a century-long head start in articulating how genetics should be used to feed into hierarchical ideology and oppressive policies, so we anti-eugenicists have our work cut out for us. In this final chapter, then, I hope to start the conversation about what it means for science and policy to be actively anti-eugenicist, by offering five general principles:

1. Stop wasting time, money, talent, and tools that could be used to improve people's lives.
2. Use genetic information to improve opportunity, not classify people.
3. Use genetic information for equity, not exclusion.
4. Don't mistake being lucky for being good.
5. Consider what you would do if you didn't know who you would be.

For each of these principles, I will contrast three positions. First, the *eugenic* position positions genetic influence as a naturalizer of inequality. If social inequalities have genetic causes, then those inequalities are portrayed as the inevitable manifestations of a "natural" order. Genetic information about people can be used to slot them more effectively into that order. Second, the *genome-blind*[4] position sees genetic data as the enemy of social equality and so objects to any use of genetic information in social science and policy. Whenever possible, the genome-blind position seeks *not to know*: scientists ought not to study genetic differences or how they are linked to social inequalities, and other people in society ought not to use any scientific information that is generated for any practical purposes. These two positions can be contrasted with what I am proposing is an *anti-eugenic* position that does not discourage genetic knowledge but deliberately aims to use genetic science in ways that reduce inequalities in the distribution of freedoms, resources, and welfare.

Stop Wasting Time, Money, Talent, and Tools

> EUGENIC: Point to the existence of genetic influence to deny
> the possibility of intervening to improve people's lives.
> GENOME-BLIND: Ignore genetic differences even if it wastes
> resources and slows down science.
> ANTI-EUGENIC: Use genetic data to accelerate the search
> for effective interventions that improve people's lives and
> reduce inequality of outcome.

Everything is heritable. This stylized fact was proposed as the "first law of behavioral genetics"[5] by Eric Turkheimer two decades ago. And Turkheimer was formulizing what had been suspected to be true for decades before him. Theodosius Dobzhansky, the famed evolutionary biologist, is worth quoting once more: "People vary in ability, energy, health, character, and other socially important traits, and there is good, although not absolutely conclusive, evidence that the variance of all these traits is in part genetically conditioned. Conditioned, mind you, not fixed or predestined."[6]

The fact that income, educational attainment, subjective well-being, psychiatric disease, neighborhood advantage, cognitive test performance, executive function, grit, motivation, and curiosity are all heritable does *not* mean that these things cannot be improved by intervention or bolstered by environmental privilege. They can.

But the fact that these things are heritable does mean that vast amounts of research in the social sciences, designed to identify specific environments that could be targeted with new interventions, *waste time and money*. The studies that I have in mind are a waste because their research designs depend on correlating some aspects of a person's behavior or functioning with some aspect of the environment that is provided by a biological relative, such as a parent, without controlling for the fact that biological relatives can be expected to resemble each other *just* because they share genes. This methodological flaw would perhaps be excusable if these fields had a track record of rapid progress in the development of successful intervention programs to improve children's lives. But they don't.

Opportunity cost is real. We do not live in a world where there is unlimited time and research funding and trained scientific talent and political will to intervene. Getting it wrong has consequences, in terms of money and effort not devoted elsewhere. Deliberately risking the possibility of getting it wrong, again and again, in an entirely predictable way, by failing to even *consider* the role that genetics plays in how children's lives turn out differently, is egregious in its wastefulness.

The anti-eugenic scientist and policymaker is concerned with mitigating inequalities, including inequalities in health and well-being that are caused by differences between people in their genetic risk. This goal requires developing effective interventions to improve people's lives. As I described in detail in chapter 9, genetic data can be a crucial tool for this endeavor, by improving our basic science about how specific environments cause specific outcomes, and by helping us assess whether interventions are serving the needs of people who are most "genetically" at risk. (I have put "genetically" in quotes here, because, as I've explained throughout the book, genetic differences between people might be connected to outcomes via social mechanisms, yet measuring DNA allows researchers to see a dimension of risk that might otherwise go unobserved.)

If social scientists are going to rise to the challenge of actually improving people's lives, they can no longer engage in the "tacit collusion" to ignore a key source of why people's lives turn out differently—their DNA.

Use Genetic Information to Improve Opportunity, Not Classify People

Eugenic: Classify people into social roles or positions based on their genetics.

Genome-Blind: Pretend that all people have an equal likelihood of achieving all social roles or positions after taking into account their environment.

Anti-Eugenic: Use genetic data to maximize the real capabilities of people to achieve social roles and positions.

"Everyone will know who you are, what you are about. To me that is really scary. . . . A world where people are slotted according to their inborn ability—well, that is *Gattaca*."[7] This was the dark prognostication of the sociologist Catherine Bliss, in an interview with *MIT Technology Review* on the growing availability of polygenic indices related to socially valued outcomes like educational attainment or criminal behavior. *Gattaca*, of course, is the 1997 movie, cleverly titled using only the letters for DNA base pairs, and starring Ethan Hawke as an aspiring astronaut grounded by his status as a genetic "in-valid." The movie led to the marriage of Hawke and co-star Uma Thurman, and to countless questions from undergraduates, seminar audiences, and journalists about whether the latest study in behavioral genetics is going to lead to a dystopian society.

While no one has yet proposed labeling children, *Gattaca*-style, as genetically "in-valid" based on their polygenic scores, there have been suggestions by several prominent behavioral geneticists to use polygenic scores for selection in education and occupational contexts. Foremost among them is Robert Plomin, a psychologist and behavioral geneticist who has had a long and illustrious career conducting both twin studies and studies using polygenic indices. In his book *Blueprint*, for example, he proposed that "a password-protected link to a direct-to-consumer company could make available a certified set of polygenic scores relevant to occupational selection in general and different sets of polygenic scores relevant to different jobs."[8] Such proposals to select people for desirable educational and occupational positions based on their measured DNA are flawed on both empirical and moral grounds.

Empirically, we must grapple with the effect sizes of polygenic scores for individuals. In the context of social science research, polygenic scores can be incredibly useful, because the ability of a DNA-based variable to capture 10 percent of the variance in a complicated outcome like educational attainment rivals the effect sizes of other variables that social scientists commonly use, such as family income. This type of research draws conclusions about the *average* outcomes seen for people who have low versus high polygenic scores. It is considerably easier to make predictions about averages in groups of people than it is to predict the outcome for a single individual, precisely because averages do "average out" all the idiosyncratic and serendipitous events that make an individual life unpredictable. The sorts of tests that we use to diagnose something about an individual— an at-home pregnancy test, for instance, or the lab test that your doctor might use to diagnose you with strep throat—are much more accurate for individual prediction than any polygenic index. And that is *before* you consider all the other sources of information we have about an individual person in a selection context—their previous grades and test scores and work histories, for example.[9]

But even if a polygenic index were much more accurate in its predictions for an individual, it is *still* problematic to assign people to social roles and positions on the basis of their measured genotype.

Let's return to the example of the cookbook-wide association study that I explained in chapter 3: you assembled a dataset that has Yelp ratings of every restaurant in town, and you correlated those Yelp ratings with bits and bobs from each restaurant's set of recipes, generating a set of small correlations between recipe elements ("add cumin") and higher-rated restaurants. You could then use those correlations to create a "poly-culinary index" for a new restaurant: analyze their planned menu and score it based on whether the new restaurant's recipes have more of the elements that are correlated with being rated highly on Yelp. This situation is analogous, of course, to conducting a GWAS and creating a polygenic index.

Now, imagine that it becomes de rigueur for investors to calculate the poly-culinary index of proposed new restaurants, and only restaurants above a certain threshold are able to raise sufficient funding. This practice creates a *feedback loop*—qualities that are statistically associated with one metric of success, at one time and place, become even more associated with success, because restaurants with those qualities are rewarded with opportunities and investment that other restaurants don't get.

This feedback loop is key to creating what have been called "weapons of math destruction" or "algorithms of oppression."[10] Already, many industries use predictive tools to automate treating particular people in particular ways. Instagram and Google display ads targeted based on your demographics and social media activity and web searches and purchase history. Mortgage lenders set interest rates based on automated algorithms that predict repaying one's loan. Police departments use data on past crime, neighborhood features such as bars, schools, and take-out restaurants, and even the weather to target communities with increased police surveillance. And once a person has contact with the criminal justice system, automated risk assessments are used in decisions about bond, sentencing, and parole.[11] These apparently objective and neutral algorithms can entrench social inequalities.

An excellent example is a commercial risk-prediction algorithm used by large health care systems to identify patients for "high-risk care management programs," which are expensive and have scarce availability. A revealing 2020 study in *Science* compared patients who received the same algorithmic risk score but who differed in their self-identified race. It found that, at any given score, patients who self-identified as Black were much sicker than patients who self-identified as White. The problem with the algorithm stemmed from the fact that Black people have, on average, worse access to health care and receive less of it, which means that less money is spent on them. The algorithm, however, uses money spent on previous care as if it were an unbiased indicator of someone's health, leading to the under-recognition of Black patients who could benefit from high-risk care management. The institutionalized racism that leads to racial disparities in health care was codified into the algorithm, which then led to fewer Black people getting the additional medical help they needed. In this way, "technoscience reflects and reproduces social hierarchies, whether wittingly or not."[12]

Like other predictive algorithms, polygenic indices use information about the past to make predictions about the future. A polygenic index that predicts educational attainment or academic achievement or occupational success is picking up on *any* heritable characteristic that was correlated with these outcomes in the samples of people who were studied, as well as any characteristic of people's *parents* that was correlated with their children's outcomes in these samples. As a result, a polygenic index, when used to classify people, is as vulnerable as any other predictive algorithm to reproducing social hierarchies—including ones that we would recognize as patently unfair if not masked by the apparent neutrality of DNA.

For example, we would consider it unfair to measure family income just so that we can deny university admission to low-income students on the grounds that they are less likely to graduate from college. Regardless of its predictive ability, parental socioeconomic status is a characteristic over which students have no control or agency. The relationship between family income and college completion is a problem to be solved, an inequality to be closed, not a result to be leveraged to further exclude low-income students. But, as I described in chapter 9, studies of parent-offspring trios have shown that part of what a person's polygenic index is picking up on is environmental advantages that are correlated with their parents' genes. Selecting a student on measured DNA is, in part, selecting a student based on their family's socioeconomic status.

Unfortunately, this danger has been actively minimized by many scholars talking about polygenic scores. In *Blueprint*, for instance, Robert Plomin claimed that polygenic indices are particularly useful for educational and occupational selection, because they are "more objective and free of biases like faking and training . . . You can't fake or train your DNA." The conservative writer Charles Murray made a similar claim in an op-ed for the *Wall Street Journal:*[13] that polygenic indices are "impervious to racism and other forms of prejudice." This is simply not true. A GWAS will pick up on *any* genes associated with educational outcomes regardless of what social mechanisms are responsible for creating that association. These social mechanisms might include mechanisms that we consider acceptable (e.g., children who are more interested in school go further in school), but also mechanisms that are more controversial and arbitrary (e.g., children who are morning people go further in school). Creating a polygenic index based on those GWAS results and using it to assign people to social roles, then, will codify these arbitrary and controversial processes, rendering them invisible under a guise of "objective" prediction.

Given these concerns, how might polygenic scores be used more productively? Let's go back to a specific example that I told you about in chapter 7, about the relationship between the educational attainment polygenic index and mathematics course-taking in high school. Students who had a higher polygenic index were more likely to be enrolled in geometry (versus algebra 1) in the ninth grade, which put them on track to complete calculus by the end of high school. Students who had a higher polygenic index were also less likely to drop out of math once it became optional. What can and should be done with that information?

The eugenic proposal would be to test students' DNA and use it to assign them to mathematics tracks, such that students with low polygenic indices are excluded from opportunities to learn advanced mathematics. The gene-blind proposal would be to insist that the research connecting genetics and mathematics course taking shouldn't have been done in the first place. The anti-eugenic proposal is to apply that genetic knowledge toward (a) understanding how teachers and schools can maximize the mathematics learning of their students, and (b) spotlighting how academic tracking entrenches inequalities between students.

Regarding the first goal, consider that one of the greatest challenges to understanding which teachers and schools are best serving the needs of students is that students with different learning needs are not randomly distributed across teachers and schools. A trenchant criticism of using standardized test scores as a metric for teacher and school "accountability"—that is, for identifying poorly performing teachers and schools—is that student test scores are highly correlated with student characteristics, such as family socioeconomic status, that precede the child's entry to school and that are non-randomly clustered across schools.[14] "Good" schools, defined as schools with high average test scores, are, in actuality, often better described as rich schools with high concentrations of affluent students. (A similar problem besieges identifying the best doctors and hospitals: the best doctor is not the one who avoids treating the sickest patients.)

Researchers have long recognized that estimating school effects on student academic outcomes is a tricky problem,[15] and one can begin to make fair, "apples-to-apples" comparisons among schools only if one incorporates measures of student characteristics such as family background, previous levels of academic knowledge, etc. The appropriate question is not "How do students in school X fare differently than students in school Y?" because the students in school X could be already different from the students in school Y in ways *other* than the school they attend. The appropriate question is, "How would a particular student have fared differently if he had attended school X rather than school Y?" (Again, we see the importance of counterfactual reasoning for causal inference, as I explained in chapter 5).

In attempting to identify school effects, it is commonplace for researchers, educators, and policymakers to consider information about one accident of birth: a student's socioeconomic status. But I and others have observed in our research that information from a student's DNA, in the form of a polygenic index, also predicts academic outcomes, above and beyond information on family socioeconomic status. As I described above, this does not mean that we should use polygenic indices to classify students and restrict their opportunities to learn. It does mean, however, that we can evaluate how students who have equivalent polygenic indices fare differently in their outcomes when they attend different schools.

In one study of US high school students, we found that students with low education-related polygenic indices were, on average, less likely to continue in their mathematics education in high school. But their dropout rates differ substantially across school contexts. In schools that primarily serve students whose parents have high school diplomas, even students with low polygenic indices take a few years of math after the ninth grade. In fact, students with low polygenic indices in high-status schools fare about as well, in terms of their persistence in math, as students with average polygenic indices who attend low-status schools.[16]

This finding is just barely scratching the surface. What, specifically, is happening in higher-status schools that keeps even students who are statistically likely to drop out of math from actually dropping out? How do you make the practices of such schools more widely available to all students? The path from basic research like this study to educational policy reform is long and tortuous.

But even though it is just a first step, this study is revealing a basic and important truth: given a certain fixed starting point in life—inheriting a certain combination of DNA variants—some people get much further in developing their capability to solve mathematical problems. These mathematical skills have lifelong benefits for an individual in terms of future education, participation in the labor force, and ease with navigating problems of everyday living. In fact, math literacy is so important for a student's future that the opportunity to learn math has been called a civil right.[17] Genetic data has thus revealed an inequality of environmental opportunity, one that calls out for redress.

Other environmental inequalities could be similarly diagnosed using genetic data. Which health interventions reach people who are currently most genetically at risk for poor outcomes? Which schools have the lowest rates of disciplinary problems among youth who are currently at most genetic risk for aggression, delinquency, or substance use problems? Which areas of the country are "opportunity zones," where opportunity is defined not solely in terms of how children from low-income families fare, but also in terms of how children who are genetically at risk for school problems or mental health problems fare? If researchers embrace principle #1, and start embracing the possibilities of genetic data, we will have a wealth of new information to address these questions.

Use Genetic Information for Equity, Not Exclusion

EUGENIC: Use genetic information to exclude people from health care systems, insurance markets, etc.

GENOME-BLIND: Prohibit the use of genetic information per se but otherwise keep markets and systems the same.

ANTI-EUGENIC: Create health care, educational, housing, lending, and insurance systems where *everyone* is included, regardless of the outcome of the genetic lottery.

Vaucluse is a fancy French restaurant on the Upper East Side of Manhattan that evokes adjectives like "posh" and "gilded." Along with a handful of other academic scientists, I ate dinner there one autumn evening as the guest of a billionaire philanthropist who made his considerable fortune in the insurance business. The conversation was lively, with all of us eagerly debating how to interpret new advances in the field of behavioral genetics. But any sense that the conversation was purely academic dissipated quickly when our host let out a sharp laugh and commented: As an insurance executive, why wouldn't he use genetics to make money?

What he meant, of course, is that genetic discoveries and the creation of polygenic indices could be used to improve predictions about people's risks for bad outcomes. And if people with high risk are charged more in premiums, or denied coverage altogether, then profits could increase. But whereas a billionaire insurance executive sees genetic prediction as an open door to making even more profit, many ordinary Americans might fear that genetic prediction opens the door to financial ruin. Health care costs, including the cost of insurance premiums, deductibles, and uncovered medical bills, are already the leading cause of bankruptcy in the United States.[18] What if you lost your coverage, or your premiums increased, because your insurer knew something about your genome?

It was precisely this fear that motivated the passage of the Genetic Information Nondiscrimination Act (GINA), which was signed into law in 2008 after a "glacial" pace of Congressional deliberations. GINA prohibits genetic information being used for discrimination in health insurance and employment, in order to "fully protect the public from discrimination and to allay their concerns about the potential for discrimination, thereby allowing individuals to take advantage of genetic testing, technology, research and new therapies."[19] GINA epitomizes a genome-blind approach, in that employers and insurers are prohibited from requiring or using genomic information. Decisions are to be made as if that information doesn't exist or—poof!—has been made to disappear.

Despite the lofty goal of "fully" protecting the public, GINA has notable limitations. First, its protections only apply to health insurance and employment, but not to other forms of insurance, like long-term care insurance, life insurance, or mortgage insurance, and not to educational contexts or housing or lending. A review of the impact of the legislation after ten years found that, although GINA may have "important symbolic value," its actual practical value has been limited.[20] The first part of the act, dealing with health insurance, is "largely irrelevant," surpassed by the Affordable Care Act, which prohibits insurers using health status when making underwriting decisions, while the second part of the act, dealing with employment, is rarely invoked. Some of these limitations have been addressed at the level of individual US states. California, in particular, is notable for passing the California Genetic Information Nondiscrimination Act (CalGINA), which is broader in scope, prohibiting discrimination based on genetic information not just in health insurance and employment, but also in housing, education, mortgage lending, and public accommodations.

Within the framework of anti-discrimination law, GINA (and genome blindness generally) is an "anti-classification" approach, in that genetic information is, like race or religion, a "forbidden" characteristic that can't be used as the basis for intentionally different treatment.[21] Anti-classification approaches to discrimination law operate under a "sameness" model of civil rights: people who could be differentiated according to some characteristic (Black versus White, male versus female, Christian versus Jewish, carrier of *APOE* ε-4 allele vs. carrier of *APOE* ε-3 allele) must be formally treated the same.[22]

The legal scholar Mark Rothstein has pointed out that the genome-blind, anti-classification approach of GINA is severely challenged by the difficulty of cordoning off "genetic" information from the medical and behavioral information that genomic data can predict.[23] Before the Affordable Care Act (ACA) provided protections to people with pre-existing conditions, insurers would *not* be able to discriminate against someone who, for example, had a *BRCA* mutation, but had not yet developed breast cancer. The moment she did develop breast cancer, however, she would be vulnerable to increased premiums or dropped coverage. But those ACA protections for pre-existing conditions were only workable in combination with an individual mandate to purchase insurance. Without it, the pool of people with insurance would contain too few low-risk people to be economically sustainable. *But*, the ACA mandate to buy insurance has been politically controversial, to say the least, motivating the rise of the ultra-right Tea Party wing of the Republican party and (as of early 2020) only narrowly surviving challenges to its constitutionality. As Rothstein wryly asked, "Is it possible to prevent genetic-based discrimination in health insurance within a system that is unfair and illogical? Unfortunately, the answer is no, unless and until the United States is prepared to address in a comprehensive way the larger issue of who has access to health care."[24]

The alternative to the "anti-classification" approach to discrimination law is the "anti-subordination" approach, which focuses on raising the social status of certain marginalized or oppressed groups and preventing the formation of an underclass.[25] In contrast to anti-classification, which forbids differential treatment, anti-subordination allows for *positive* differential treatment. The Individuals with Disabilities Education Act (IDEA), for instance, takes an anti-subordination rather than anti-classification approach. Under IDEA, children are held to have an equal right to a "free, appropriate public education." In designing an appropriate education, school systems are not only *allowed* to consider certain differentiating information about the individual student; they are, in fact, *mandated* to consider that information for accommodation and planning purposes.

This principle of anti-subordination is critical to formulating anti-eugenic policy, not just in the areas of health insurance and education, but also in other forms of insurance, employment, lending, and housing. Eugenic policy, historically and in the present day, works to create and subjugate an economic and racial underclass by labeling people in that underclass as biologically inferior. Anti-eugenic policy, then, must fight the emergence of a new "genetic" underclass, i.e., where people are excluded from access to health care, housing, lending, or insurance on the basis of traits, such as their health or educational history, that are themselves partly the outcome of the genetic lottery. In the realm of health care, for instance, rather than a genome-blind approach that narrowly forbids the use of genetic information yet keeps everything else about the American health care system the same, a more fully anti-eugenic response to the rising tide of genetic discoveries is a commitment to truly universal health care, to which *everyone* has access, regardless of the outcome of the genetic lottery (*or* the environmental one).

Don't Mistake Being Lucky for Being Good

EUGENIC: Point to genetic effects on intelligence as proof that some people naturally have more merit than others.

GENOME-BLIND: Accept the logic of meritocracy while ignoring the role of genetic luck in developing skills and behaviors that are perceived as meritorious.

ANTI-EUGENIC: Recognize genetics as a type of luck in life outcomes, undermining the meritocratic logic that people deserve their successes and failures on the basis of succeeding in school.

America, we are told, is a meritocracy. The word is a portmanteau of *merit* and *aristocracy*. Aristocracy, in turn, is from the Greek *aristokratia—aristos* meaning best, and *kratos* meaning rule. Embedded in the term "meritocracy," therefore, is the idea that elites in a society, i.e., people selected for positions of power, influence, wealth, and prestige, should be those selected on their merits. When compared to a rigid class or caste system, where the schools you could attend and the jobs you could take and the roles you could play in public life were strictly gated by the station of your birth, the idea of meritocracy has, well, merit. My father grew up in a Texas trailer park and became an officer in the US Navy; none of my grandparents went to college, whereas I have a PhD. Reciting these "American Dream" success stories nurtures and sustains the mythology that anyone can succeed in this country, regardless of our origins.

When people criticize the idea of meritocracy, they usually are arguing that America is *not meritocratic enough*. The college admissions scandal of 2019, when Hollywood actresses and other wealthy parents were arrested for bribing athletic coaches and faking test scores in order to get their children admitted to elite universities, was a tragicomic example of how the upper social classes can reproduce themselves in an ostensibly meritocratic competition—by outright cheating. Even without outright lies and bribes, rich students with low scores on their SAT college admissions tests are *still* more likely to graduate from college than poor students with high scores. These stories and statistics reveal that American society is far from the meritocratic ideal.

But even if we were entirely successful at eliminating inequalities of outcome associated with being born into wealth or privilege, the inequalities that remain would not be purged of luck. There would still be another type of luck lurking in the background: genes. This is true not only of standardized test performance and IQ scores. Even appealing to so-called "character" traits (grit, perseverance, resourcefulness, motivation, curiosity, or any other non-cognitive skill) doesn't get you out of grappling with genetics. These traits, too, are shaped by genetic differences between people. There is no measure of so-called "merit" that is somehow free of genetic influence or untethered from biology.

In light of the ubiquity of genetic influence, "merit" is a deeply misleading word for the skills and behaviors that are currently associated with educational and economic success. Consider the ordinary dictionary definition of "merit":

1. a) (obsolete) reward or punishment due;
   b) the qualities of actions that constitute the basis of one's deserts;
   c) a praiseworthy quality: virtue;
   d) character or conduct deserving reward, honor, or esteem; achievement;
2. spiritual credit held to be earned by performance of righteous acts and to ensure future benefits.

"Virtue." "Spiritual credit." "Righteous acts." "Character." "Conduct deserving reward." The word "merit," in ordinary usage, has a distinctly moral tone. And, in casually using a word that connotes moral deservingness to describe the skills and behaviors that are used to select a person for a socially desirable role, we risk conflating these skills and behaviors with human character and worth.

Connecting people's biology to their virtue, righteousness, and moral deservingness is a eugenic idea. To say that some people *deserve* more power, more resources, more freedom, more welfare because they were born with certain genotypes *is* inegalitarian.

The gene-blind response, however, simply accepts the logic of meritocracy, where some people are held to deserve more because of their "merit," without grappling with the role of genetic luck in producing the differences between people that we've labeled as meritorious. Gene blindness thus perpetuates the myth that those of us who have "succeeded" in twenty-first-century capitalism have done so primarily because of our own hard work and effort, and not because we happened to be the beneficiaries of accidents of birth—both environmental *and* genetic.

The appropriate response to eugenics, then, is not to avoid any discussion of genes, but rather to break up with the idea that America is or *could ever be* the sort of "meritocracy" where social goods are divided up according to what people deserve. There is no way to purge luck from human affairs; no way to disentangle, particularly for any one person, how much she deserves by virtue of her character and resourcefulness from how much she happened to benefit from a constellation of genetic and environmental advantages. As Rawls wrote, "None of the precepts of justice aims at rewarding virtue. . . . The idea of rewarding desert is impracticable."[26]

Humans recoil at the recognizing the role of luck in their lives. When the economist Robert Frank described the importance of external luck for people's economic success, a Fox News host responded with outrage: "Do you know how insulting that was?" The encounter proved the truth of the E. B. White quote that Frank used as an epigraph for his book *Success and Luck*: "Luck is not something you can mention in the presence of self-made men."[27]

But reluctant though we might be, coming to grips with the role of luck in our lives—including the role of genetic luck—is vital to the egalitarian project. As the writer David Roberts argued:[28]

Individually, coming to terms with luck is the secular equivalent of religious awakening, the first step in building any coherent universalist moral perspective. Socially, acknowledging the role of luck lays a moral foundation for humane economic, housing, and carceral policy.

Building a more compassionate society means reminding ourselves of luck, and of the gratitude and obligations it entails, against inevitable resistance.

Recognizing the role of luck, both genetic and environmental, in shaping the development of socially valued skills and behaviors does not mean that we should abandon using selection criteria for desirable social roles and opportunities. Consider, for example, a piloting job. "Meritorious" applicants for a pilot's jobs are those who can fly a plane in nasty weather without crashing it; who will reliably show up for work so millions of dollars aren't lost as a fully laden plane is left sitting pilotless on the runway; who have good eyesight and good manual dexterity and good spatial rotation abilities, and who don't have narcolepsy, and who aren't too tall to fit comfortably in a cockpit.

When one considers aviation or another sort of high-stakes profession, where failure can kill people, choosing applicants on their "merits" has obvious benefits for everyone in society. We want pilots to be selected for their ability to fly planes, not their social connections. We want surgeons, engineers, pharmacists, teachers, plumbers, etc., to be people who will operate, build, dispense, teach, and fix skillfully.

But even as we recognize that it is instrumentally *useful* to select pilots based on attributes such as good eyesight and spatial rotation skills, we can simultaneously recognize that those attributes, and the financial rewards that follow from having them, are not a sign of the pilot's moral creditworthiness or virtue. Having those attributes, *in combination* with living in a time and place where those skills can be put to economically valuable use in the form of flying planes, is like winning the Powerball. A lot of lucky events had to come together in a person's life in order for those attributes to develop and be remunerated. Good eyesight is "meritorious" in what the economist and philosopher Amartya Sen called a "derivative and contingent way, depending on . . . the good that can be brought about by rewarding [it]."[29] It is not "right conduct" that is separated out for "praise and emulation—independent of the goodness of the consequences generated."

This is obvious in the case of eyesight, but the point is often obscured in the case of cognitive ability, and even more obscured in the case of non-cognitive skills such as self-regulation or intellectual curiosity. When selecting students for scarce educational opportunities, or selecting employees for desirable jobs, selection on the basis of certain cognitive skills might be instrumentally useful for society as a whole. But the possession of those cognitive skills is no more virtuous, no more inherently deserving of reward, than having 20/20 vision. As Madeline L'Engle put it in *A Wrinkle in Time*, her classic young adult novel, "But of course we can't take any credit for our talents. It's how we use them that counts."[30]

And, if merit is defined instrumentally, then our definitions of merit cannot be separated from our definition of *what constitutes a good society*. What is considered "meritorious" is simply what brings about the social consequences that we desire. These desirable social consequences include the efficient allocation of scarce opportunities to the people who are most likely to profit from them, and the allocation of jobs to the people who are most likely to do them well. But as Amartya Sen pointed out in his essay on merit, these are not the *only* social consequences that might be desirable. We might also conceptualize a good society as one that does not have gaping economic inequalities and that does not allow the members of one racial group to dominate all elite institutions. As a consequence, Sen writes, when we are assessing "what counts as merit," we are compelled to take into stock whether the rewarding of that sort of merit mitigates or exacerbates the economic inequalities or racial disparities we care about: "The rewarding of merit cannot be done independent of its distributive consequences."

The appropriate characterization of merit is debated at every stage of the educational system, often crystallizing around the role of standardized tests. Should elite public high schools in New York City continue to use a single standardized test score as the criterion for admission, even though this admissions process leads to a severe under-representation of Black students? Should the Graduate

Record Examination be required for admission to PhD programs? What Sen is saying here is that the *only* criterion for what makes something a "good" definition of merit is the *consequences* of rewarding merit according to that definition. If, for instance, the racial inequalities that result from admitting students on the basis of a single standardized test are unacceptable (and not countervailed by other benefits to society), then that test is not an instrumentally good definition of "merit."

Thus, rather than naturalizing inequalities as the result of inherent differences between people in how much they deserve, a serious consideration of the role of genetics in cognitive ability, non-cognitive skills, and social inequalities generally, combined with a commitment to anti-eugenics, leads us to a very different perspective: None of us deserves his or her genetics. To the extent that we enjoy good things in life—educational success, good incomes, stable jobs, good physical health, happiness and subjective well-being—it is, in large part, because we have been massively lucky. The ubiquity of genetic influence on human individual differences renders it impossible to construct an educational or economic system that purely rewards people for what they morally deserve. "Merit" in the meritocracy is thus a hollow concept that can only be defined instrumentally, in terms of how selecting on a particular set of criteria brings about *the sort of society that we want to live in.*

Consider What You Would Do, If You Didn't Know Who You Would Be

EUGENIC: The biologically superior are entitled to greater freedoms and resources.

GENOME-BLIND: Society should be structured as if everyone is exactly the same in their biology.

ANTI-EUGENIC: Society should be structured to work to the advantage of people who were least advantaged in the genetic lottery.

The coffee shop near my house sells giant chocolate chip cookies. On summer afternoons, I'll walk there with my children, and we'll buy one cookie to share. The rule is that one child can choose how the cookie is divided, but then the pieces of cookie are hidden behind my back and they have to pick randomly. Consistent with children's strong preferences for equality, my kids always choose to divide the cookie as equally as possible.

How do you divide the cookie if you don't know which piece you'll get? It's a premise that will be familiar to anyone who has taken an undergraduate class in political philosophy. The most famous version of this thought experiment was proposed by the philosopher John Rawls, who imagined something called "the veil of ignorance." Behind the veil of ignorance,[31]

> no one knows his place in society, his class position or social status; nor does he know his fortune in the distribution of natural assets and abilities, his intelligence and strength, and the like. Nor, again, does anyone know his conception of the good, the particulars of his rational plan of life, or even the special features of his psychology such as his aversion to risk or liability to optimism or pessimism.

The point of the veil of ignorance is to imagine a hypothetical situation in which everyone is on an equal footing and so can come to a fair agreement about the principles of justice: if you didn't know who you were going to be, and neither did anyone else, and you had to decide on the basic structure of society while radically ignorant about the particulars of your own self-interest, what would be the rules for deciding who gets to do what and who gets to have what?

Rawls argued that two principles would emerge from the fair agreement of people behind the veil of ignorance:

1. Each person has an equal right to a fully adequate scheme of equal basic liberties which is compatible with a similar scheme of liberties for all;
2. Social and economic inequalities are to be (a) attached to offices and positions open to all under conditions of fair equality of opportunity, and (b) they must be to the greatest benefit of the least advantaged members of society.

When considering whether inequalities are to everyone's advantage, Rawls does not mean simply that the average is improved; if the already-disadvantaged are made worse off by inequalities, but the advantaged are made even better off, that isn't good enough. Inequalities must be arranged so that they work to the advantage of the *least* well-off person. Rawls explains, "Those who have been favored by nature, whoever they are, may gain from their good fortune only on terms that improve the situation of those who have lost out."

We can also look to recent history[32] to see enormous gains in life span, literacy, wealth, and well-being that ultimately worked to everyone's advantage—such that, in Rawls's words, the "distribution . . . of natural talents" worked "as in some respects a common asset" to produce social and economic benefits. Between 1820 and 1992, the average income of the world's people grew 8 times bigger, while the share of people in extreme poverty fell from 84 percent to 24 percent.[33] In 1700s Sweden, one in three babies died before their fifth birthday, a scale of infant death that is nearly impossible for me to comprehend emotionally. Today in Sweden, the infant mortality rate is two in a thousand—more than 100 times lower.[34]

The innovations in science, technology, and government that improved people's lives were inequality-producing: some people's lives were made better, quicker, than others and these innovations, in some cases, were inequality-dependent, in that they were made possible by a system that differentially rewarded different types of skills.[35] But it's to everyone's advantage to live in a society where we don't lose one-third of our children. As I described in the previous section on merit, rewarding certain skills might be instrumentally useful for society as a whole, even as we recognized that people didn't *deserve* the fact that they inherited genetic variants that were among the causes of those skills.

But notice how different this justification for inequality is than the one we often encounter in our so-called "meritocracy." As Rawls explains, the second principle "transforms the aims of the basic structure so that the total scheme of institutions no longer emphasizes social efficiency and technocratic values. . . . The naturally advantaged are not to gain merely because they are more gifted, but only to cover the costs of training and education and for using their endowments in ways that help the less fortunate as well." Allocating certain educational opportunities to certain people on the grounds that this allocation is most likely to benefit everyone is different from asking who "deserves" to go to Harvard.[36]

The research that I've described in this book makes the case that the "naturally advantaged"—a.k.a. people who have happened to inherit certain genetic variants—do have better life outcomes, in terms of their educational success and income and wealth and well-being. These genetically caused inequalities are not fixed and inevitable outcomes of natural law, but a function of how genetic differences between people in their cognitive abilities, personality traits, and other personal characteristics are refracted through the prism of our economy and social institutions. The key question raised by Rawls's principles of justice is: Are these inequalities working to the advantage of *everyone*, even those who are least well-off in the distribution of genetic variants currently associated with success?

As I was writing this chapter, I had coffee with my former neighbor, whose girlfriend had just died of sepsis after she fell and hit her head. She was an alcoholic who had been in and out of rehab for the better part of a decade, never being able to sustain sobriety for very long. She was in her early fifties, and her death is just one of what the economists Anne Case and Angus Deaton have called "deaths of despair"—deaths due to suicide, drug overdose, and alcoholism that disproportionately affect Americans without a college degree.

This disturbing and historically unprecedented rise in mortality is just the tip of an iceberg of poor health, mental anguish, financial precarity, frayed family relationships, and chaotic living situations. Case and Deaton conclude that, while capitalism lifted millions of people out of poverty and poor health from the late 1700s to the late 1900s, it has now become toxic, producing inequalities that are not justifiable in terms of being for the collective advantage.[37] Our prosperity as a nation is not being broadly shared.

Take the power of the genetic lottery seriously, and you might be faced with the realization that many of the things you pride yourself on, your high vocabulary and your quick processing speed, your orderliness and your "grit," the fact that you always did well in school, are the consequence of a series of lucky breaks for which you can take no credit. Now, take the Rawlsian thought experiment about the veil of ignorance seriously, and consider: What sort of society would you want if you didn't know what the outcome of the genetic lottery was going to be?

Conclusion

As I write the last chapter of this book, my university and my children's school have shut down, in an attempt to slow the transmission of the COVID-19 virus. Sarah Bessey, an author and pastor, summarized the various recommendations given by public health officials as all being versions of the same message: "Love the vulnerable with your choices." As a physically healthy woman in her thirties, with good access to medical care in a highly resourced urban area, I might not be particularly vulnerable to serious illness from coronavirus. But my elderly neighbor is, and her situation will be additionally worsened if the capacity of the medical system is overly taxed by too many people getting sick in the same way in the same place at the same time.

Responding to the threat of pandemic illness involves us defining our responsibilities to each other in order to protect the most vulnerable among us. Our responsibilities to each other include individual behavioral change (e.g., washing hands and wearing masks) and also effective institutional response (e.g., financial relief measures to alleviate the pressures that would otherwise induce sick people to go to work). What our responsibilities to each other do *not* include, however, is pretending that everyone is equally vulnerable to disease. In fact, insisting that everyone's biological vulnerability to COVID-19 is the same—young or old, immune-compromised or healthy—would be both ludicrous and dangerous. Protecting the most vulnerable requires knowing who is most vulnerable, identifying what factors make them most vulnerable, and structuring society for their benefit.

But our responsibilities to each other don't end once the threat of pandemic disease is lifted. Right now, society is structured in such a way that only people who succeed in formal education, in particular those who have a college degree or higher, are sharing in national prosperity. People who do not have a college degree—which remains most of America—are vulnerable. Their relationships and marriages are vulnerable to dissolution; they are vulnerable to alcohol and drug abuse. They are vulnerable to anxiety and despair and suicide; they are vulnerable to crushing medical debt and unnecessary suffering due to preventable illness.

For the past century, there has been a persistent and malicious drumbeat from those espousing a eugenic ideology that the vulnerable deserve their vulnerability because of their biological inferiority. With good intentions, there has been a corresponding drumbeat from those determined to uncouple social vulnerability from biology. But a commitment to anti-eugenics does not require pretending that social vulnerability is uncoupled from biology, any more than an effective response to pandemic disease requires one to pretend that the elderly are no more susceptible than the young. A society that protects—nay, loves—its most vulnerable with its choices must be able to see who is most vulnerable, so that it can see how its choices affect them.

Some people happen to inherit combinations of genetic variants that, in combination with environments provided by parents and teachers and social institutions, cause them to be more likely to develop a suite of skills and behaviors that are currently valued in the formal education systems of Western capitalist societies. They are not better people. They are not more inherently meritorious. They are, given the ways our society is currently constructed, the least vulnerable. And, if you are reading this book, you are probably one of them.

As the threat of coronavirus ripples across the United States and the world, closing schools, shuttering businesses, socially responsible people are asking themselves: What do I need to do protect the most vulnerable in my community? This is the question we should be asking ourselves for long after the pandemic subsides.

ACKNOWLEDGMENTS

The epigraph for the book comes from *The Witch Elm* by Tana French, copyright ©2018 by Tana French, reproduced with permission of Penguin Books, an imprint of Penguin Publishing Group, a division of Penguin Random House LLC. All rights reserved.

The original idea to write a book on genetics and equality was sparked by conversations I had with scholars at the Russell Sage Foundation, where I was on sabbatical for the 2015–2016 academic year. Since then, I've had the opportunity to discuss this work and learn from my colleagues in several interdisciplinary forums, including the meetings of the Genetics and Human Agency project, organized by Eric Turkheimer and funded by the John Templeton Foundation; the Hastings Center working group "Wrestling with Social and Behavioral Genomics: Risks, Potential Benefits, and Ethical Responsibility," organized by Erik Parens and Michelle Meyer, with funding by the Robert Wood Johnson Foundation, Russell Sage Foundation, and JPB Foundation; a workshop on interpreting the genetic basis of differences between populations and on the interactions among concepts used for research in social and natural sciences, organized by Danielle Allen, Anna Di Rienzo, Evelynn Hammonds, Molly Przeworski, and Alondra Nelson, sponsored by Harvard University's Edmond J. Safra Center for Ethics; a workshop, "Genes, Schools, and Interventions That Address Educational Inequality," co-organized with David Yeager and sponsored by the Human Capital and Economic Opportunity Global Working Group at the University of Chicago; and a residency on Genes and Development, co-organized with Dan Belsky and sponsored by the Jacobs Foundation. I am indebted to all of the participants of these workshops and meetings for their incisive comments. Research for this book was further supported by grants from the Templeton Foundation and the Jacobs Foundation.

I have had the opportunity to present ideas from this book to a number of audiences, including the Duke University Population Research Institute, the Office of Population Research at Princeton University, the Department of Psychology at the University of Wisconsin, the Global Education and Skills Forum, and the Département d'Études Cognitives at the École Normale Supérieure, as well as attendees of meetings of the American Philosophical Association, Philosophy of Science Association, Behavior Genetics Association, the Integrating Genetics and the Social Sciences conference, Association for Psychological Science, American Society of Human Genetics, and American Society for Bioethics and Humanities. Thank you to these audiences for their illuminating questions and comments.

Writing a book means being distracted from one's ordinary responsibilities for a long time, and no one has borne the brunt of that distraction more than my trainees. Megan Patterson, Stephanie Savicki, Margherita Malanchini, James Madole, Laurel Raffington, Andrew Grotzinger, Travis Mallard, Aditi Sabhlok, and Peter Tanksley have been extraordinary junior colleagues whom I look forward to knowing, and working with, for years to come.

David Yeager generously facilitated a semester course release, and Jamie Pennebaker stepped in to co-teach my class, providing invaluable writing time.

Eric Turkheimer has been my mentor for nearly two decades, and nearly every page of this book bears the imprint of his influence (even if he disagrees with much of it). I also benefited from conversations with Benjamin Riley, Carl Shulman, Graham Coop, Doc Edge, John Novembre, Stuart Ritchie, Jasmin Wertz, and Razib Khan. Patrick Turley, Sanjay Srivastava, Ben Domingue, George Davey Smith and several anonymous reviewers helpfully reviewed earlier drafts of the manuscript. Alison Kalett was a careful editor and enthusiastic advocate. Innumerable people responded helpfully to my inchoate thoughts on Twitter.

I am especially thankful for the support of friends who have provided snacks, wine, advice, and encouragement throughout this long process, including Dan Belsky, Colter Mitchell, Philipp Koellinger, Nico Dosenbach, Sam Gosling, Joe Pflieger, Jane Mendle, Samantha Pinto, Jen Doleac, Sara Beckmann, and Natalia Wulfe. Every day is improved by the companionship of Travis Avery: "You and a bird flu could make me believe in fate." Micah Harden agreed to be genotyped for this book, one of many acts of brotherly love. Elliot Tucker-Drob has been, through good times and bad, my collaborator and friend; he and Barbara Wendelberger Drob are devoted co-parents, and this book would not be possible without their teamwork. Finally, I am most grateful for my children, my natural experiments in within-family genetic diversity, my most precious preoccupations, and my reasons to hope for a better world.

NOTES

Chapter 1: Introduction

1. Alex Shaw and Kristina R. Olson, "Children Discard a Resource to Avoid Inequity," *Journal of Experimental Psychology: General* 141, no. 2 (2012): 382–95, https://doi.org/10.1037/a0025907.

2. Sarah F. Brosnan and Frans B. M. de Waal, "Monkeys Reject Unequal Pay," *Nature* 425, no. 6955 (September 2003): 297–99, https://doi.org/10.1038/nature01963.

3. "Bernie's Right: 3 Billionaires Really Do Have More Wealth Than Half of America," Inequality.org, accessed July 24, 2020, https://inequality.org/great-divide/bernie-3-billionaires-more-wealth-half-america/.

4. Noah Snyder-Mackler et al., "Social Determinants of Health and Survival in Humans and Other Animals," *Science* 368, no. 6493 (May 22, 2020): eaax9553, https://doi.org/10.1126/science.aax9553.

5. Raj Chetty et al., "The Association Between Income and Life Expectancy in the United States, 2001–2014," *JAMA* 315, no. 16 (April 26, 2016): 1750–66, https://doi.org/10.1001/jama.2016.4226.

6. Laurel Raffington et al., "Analysis of Socioeconomic Disadvantage and Pace of Aging Measured in Saliva DNA Methylation of Children and Adolescents," *bioRxiv*, June 5, 2020, 134502, https://doi.org/10.1101/2020.06.04.134502.

7. Consistent with the American Psychological Association's style guidelines, I capitalize racial terms like Black and White. While there is not consensus regarding this issue, the Center for the Study of Social Policy argued that capitalizing Black "refers to not just a color but signifies a history and the racial identity of Black Americans." Moreover, they argued that "to not name 'White' as a race is, in fact, an anti-Black act which frames Whiteness as both neutral and the standard. . . . While we condemn those who capitalize 'W' for the sake of evoking violence, we intentionally capitalize 'White' in part to invite people, and ourselves, to think deeply about the ways Whiteness survives—and is supported both explicitly and implicitly." "Racial and Ethnic Identity," APA Style, accessed February 8, 2021, https://apastyle.apa.org/style-grammar-guidelines/bias-free-language/racial-ethnic-minorities; Ann Thúy Nguyễn and Maya Pendleton, "Recognizing Race in Language: Why We Capitalize 'Black' and 'White,'" Center for the Study of Social Policy, March 23, 2020,

https://cssp.org/2020/03/recognizing-race-in-language-why-we-capitalize-black-and-white/.

8. Anne Case and Angus Deaton, "Mortality and Morbidity in the 21st Century," *Brookings Papers on Economic Activity* 2017, no. 1 (2017): 397–476, https://doi.org/10.1353/eca.2017.0005.

9. Case and Deaton.

10. "The Fed—Publications: Report on the Economic Well-Being of U.S. Households (SHED)," Board of Governors of the Federal Reserve System, accessed July 24, 2020, https://www.federalreserve.gov/publications/2020-economic-well-being-of-us-households-in-2019-financial-repercussions-from-covid-19.htm; "Hispanic Women, Immigrants, Young Adults, Those with Less Education Hit Hardest by COVID-19 Job Losses," *Pew Research Center* (blog), accessed July 13, 2020, https://www.pewresearch.org/fact-tank/2020/06/09/hispanic-women-immigrants-young-adults-those-with-less-education-hit-hardest-by-covid-19-job-losses/.

11. David H. Autor, "Skills, Education, and the Rise of Earnings Inequality among the 'Other 99 Percent,'" *Science* 344, no. 6186 (May 23, 2014): 843–51, https://doi.org/10.1126/science.1251868.

12. Paul Myerscough, "Short Cuts: The Pret Buzz," *London Review of Books*, January 3, 2013, https://www.lrb.co.uk/the-paper/v35/n01/paul-myerscough/short-cuts.

13. Fredrik deBoer, *The Cult of Smart: How Our Broken Education System Perpetuates Social Injustice* (New York: All Points Books, 2020).

14. Organisation for Economic Co-operation and Development, "Education and Earnings," accessed February 3, 2021, https://stats.oecd.org/Index.aspx?DataSetCode=EAG_EARNINGS.

15. James J. Heckman and Paul A. LaFontaine, "The American High School Graduation Rate: Trends and Levels," *The Review of Economics and Statistics* 92, no. 2 (May 2010): 244–62, https://doi.org/10.1162/rest.2010.12366.

16. Jeremy Greenwood et al., "Marry Your Like: Assortative Mating and Income Inequality," *American Economic Review* 104, no. 5 (May 2014): 348–53, https://doi.org/10.1257/aer.104.5.348.

17. "Dramatic Increase in the Proportion of Births Outside of Marriage in the United States from 1990 to 2016," *Child Trends* (blog), accessed November 5, 2019, https://www.childtrends.org/publications/dramatic-increase-in-percentage-of-births-outside-marriage-among-whites-hispanics-and-women-with-higher-education-levels; T.J. Mathews and Brady E. Hamilton, "Educational Attainment of Mothers Aged 25 and Overs: United States, 2017," NCHS Data Brief (Hyattsville, MD: National Center for Health Statistics, June 10, 2019), https://www.cdc.gov/nchs/products/databriefs/db332.htm.

18. An influential paper by Kahneman and Deaton in 2010 found that the daily experience of negative emotions went down with higher household incomes, but only up to around $70,000 per year, whereas global positive evaluations of life ("my life is the best possible life for me") continued to increase with higher incomes even beyond $70,000 per year. A more recent report by Killingsworth in 2021 used a

different strategy to measure emotional experiences: participants were pinged on their smart phones and asked to report how they felt in that moment, rather than asked to report whether they had experienced a particular type of emotion the previous day. Contra to Kahneman and Deaton, Killingsworth reported that emotional well-being continued to increase with higher incomes, even among high earners. Daniel Kahneman and Angus Deaton, "High Income Improves Evaluation of Life but Not Emotional Well-Being," *Proceedings of the National Academy of Sciences* 107, no. 38 (September 21, 2010): 16489–93, https://doi.org/10.1073/pnas.1011492107; Matthew A. Killingsworth, "Experienced Well-Being Rises with Income, Even above $75,000 per Year," *Proceedings of the National Academy of Sciences* 118, no. 4 (January 26, 2021): e2016976118, https://doi.org/10.1073/pnas.2016976118.

19. Jack Pitcher, "Jeff Bezos Adds Record $13 Billion in Single Day to His Fortune," Bloomberg Quint, July 21, 2020, https://www.bloombergquint.com /markets/jeff-bezos-adds-record-13-billion-in-single-day-to-his-fortune.

20. Alicia Adamczyk, "32% of U.S. Households Missed Their July Housing Payments," CNBC, July 8, 2020, https://www.cnbc.com/2020/07/08/32-percent-of -us-households-missed-their-july-housing-payments.html.

21. Richard Arneson, "Four Conceptions of Equal Opportunity," *The Economic Journal* 128, no. 612 (July 1, 2018): F152–73, https://doi.org/10.1111/ecoj.12531.

22. Susan E. Mayer, *What Money Can't Buy: Family Income and Children's Life Chances* (Cambridge, MA: Harvard University Press, 1997).

23. Duncan, Greg J., and Richard J. Murnane, eds. *Whither Opportunity?: Rising Inequality, Schools, and Children's Life Chances* (New York: Chicago: Russell Sage Foundation, 2011).

24. James J. Lee et al., "Gene Discovery and Polygenic Prediction from a Genome-Wide Association Study of Educational Attainment in 1.1 Million Individuals," *Nature Genetics* 50, no. 8 (August 2018): 1112–21, https://doi.org/10.1038 /s41588-018-0147-3.

25. Nathaniel Comfort, "Nature Still Battles Nurture in the Haunting World of Social Genomics," *Nature* 553 (January 15, 2018): 278–80, https://doi.org/10 .1038/d41586-018-00578-5.

26. Ivar R. Hannikainen, "Ideology Between the Lines: Lay Inferences About Scientists' Values and Motives," *Social Psychological and Personality Science* 10, no. 6 (August 1, 2019): 832–41, https://doi.org/10.1177/1948550618790230.

27. Francis Galton, *Hereditary Genius: An Inquiry into Its Laws and Consequences* (London and New York: Macmillan, 1892).

28. Francis Galton, *Natural Inheritance* (New York and London: Macmillan, 1894).

29. Daniel J. Kevles, *In the Name of Eugenics: Genetics and the Uses of Human Heredity* (New York: Alfred A. Knopf, 1985; repr., Cambridge, MA: Harvard University Press, 1998).

30. Kevles.

31. Francis Galton, *Inquiries into Human Faculty and Its Development* (London: Macmillan, 1883; second edition, Macmillan, 1907, online at Project Gutenberg, http://www.gutenberg.org/ebooks/11562.

32. Kevles, *In the Name of Eugenics.*

33. Kevles.

34. Harry Hamilton Laughlin, *Eugenical Sterilization in the United States* (Chicago: Psychopathic Laboratory of the Municipal Court of Chicago, 1922), http://hdl.handle.net/2027/hvd.hc4mzw.

35. "Harry Laughlin and Eugenics: Laughlin's Model Law," a selection from the Harry H. Laughlin Papers, Truman State University, accessed November 28, 2020, https://historyofeugenics.truman.edu/altering-lives/sterilization/model-law/.

36. "Carrie Buck Revisited and Virginia's Expression of Regret for Eugenics," *Eugenics: Three Generations, No Imbeciles: Virginia, Eugenics & Buck v. Bell* (blog), accessed February 3, 2021, http://exhibits.hsl.virginia.edu/eugenics/5-epilogue/.

37. Paul Lombardo, "Three Generations, No Imbeciles: New Light on *Buck v. Bell*," *New York University Law Review* 60, no. 1 (April 1985): 30–63, https://readingroom.law.gsu.edu/cgi/viewcontent.cgi?article=2593&context=faculty_pub.

38. "DeJarnette, Joseph S. (1866–1957)," *Encyclopedia Virginia*, accessed November 28, 2020, https://www.encyclopediavirginia.org/DeJarnette_Joseph_Spencer_1866-1957#start_entry.

39. Paul A. Lombardo, "'The American Breed': Nazi Eugenics and the Origins of the Pioneer Fund," *Albany Law Review* 65, no. 3 (2002): 743–830, available at SSRN: https://papers.ssrn.com/abstract=313820.

40. Lombardo, "'The American Breed.'"

41. Lombardo.

42. "Jared Taylor," Southern Poverty Law Center, accessed November 28, 2020, https://www.splcenter.org/fighting-hate/extremist-files/individual/jared-taylor.

43. Jared Taylor, "Blueprint: How DNA Makes Us Who We Are," review, *American Renaissance*, January 4, 2019, https://www.amren.com/features/2019/01/blueprint-how-dna-makes-us-who-we-are/; Robert Plomin, *Blueprint: How DNA Makes Us Who We Are* (MIT Press, 2018).

44. Hawes Spencer and Sheryl Gay Stolberg, "White Nationalists March on University of Virginia," *The New York Times*, A12, August 11, 2017, https://www.nytimes.com/2017/08/11/us/white-nationalists-rally-charlottesville-virginia.html.

45. Richard J. Herrnstein and Charles Murray, *The Bell Curve: Intelligence and Class Structure in American Life* (New York: Free Press, 1994).

46. Richard J. Herrnstein, *I.Q. in the Meritocracy* (Boston: Little, Brown, 1973).

47. Elizabeth S. Anderson, "What Is the Point of Equality?," *Ethics* 109, no. 2 (January 1999): 287–337, https://doi.org/10.1086/233897.

48. "Remarks by the President . . . on the Completion of the First Survey of the Entire Human Genome Project," White House press release, June 26, 2000, https://clintonwhitehouse3.archives.gov/WH/New/html/genome-20000626.html.

49. J.B.S. Haldane, "KARL PEARSON, 1857–1957," *Biometrika* 44, no. 3–4 (December 1957): 303–13, https://doi.org/10.1093/biomet/44.3-4.303.

50. Roberto Mangabeira Unger, *Social Theory: Its Situation and Its Task* (Cambridge, UK: Cambridge University Press, 1987; repr., London and Brooklyn: Verso, 2004); "Roberto Mangabeira Unger's Alternative Progressive Vision," *The Nation*, July 21, 2020, https://www.thenation.com/article/culture/roberto-mangabeira-ungers-alternative-progressive-vision/.

51. Jeremy Freese, "Genetics and the Social Science Explanation of Individual Outcomes," *American Journal of Sociology* 114, suppl. S1 (2008): S1–35, https://doi .org/10.1086/592208.

52. "Susan Mayer on What Money Can't Buy," Econlib, accessed July 22, 2020, http://www.econtalk.org/susan-mayer-on-what-money-cant-buy/.

53. Jedidiah Carlson and Kelley Harris, "Quantifying and Contextualizing the Impact of *bioRxiv* Preprints through Automated Social Media Audience Segmentation," *PLOS Biology* 18, no. 9 (September 22, 2020): e3000860, https://doi.org /10.1371/journal.pbio.3000860.

54. Amy Harmon, "Why White Supremacists Are Chugging Milk (and Why Geneticists Are Alarmed)," *The New York Times*, October 17, 2018, https://www .nytimes.com/2018/10/17/us/white-supremacists-science-dna.html; Aaron Panofsky and Joan Donovan, "Genetic Ancestry Testing among White Nationalists: From Identity Repair to Citizen Science," *Social Studies of Science* 49, no. 5 (October 1, 2019): 653–81, https://doi.org/10.1177/0306312719861434; Michael Price, "'It's a Toxic Place.' How the Online World of White Nationalists Distorts Population Genetics," *Science*, May 22, 2018, https://www.sciencemag.org/news/2018 /05/it-s-toxic-place-how-online-world-white-nationalists-distorts-population -genetics.

55. Perline Demange et al., "Investigating the Genetic Architecture of Noncognitive Skills Using GWAS-by-Subtraction," Nature Genetics 53 (January 7, 2021): 35–44, https://doi.org/10.1038/s41588-020-00754-2.

56. "Pepe the Frog," Anti-Defamation League, accessed August 6, 2020, https://www.adl.org/education/references/hate-symbols/pepe-the-frog.

57. Eric Turkheimer, Kathryn Paige Harden, and Richard E. Nisbett, "Charles Murray Is Once Again Peddling Junk Science about Race and IQ," Vox, May 18, 2017, https://www.vox.com/the-big-idea/2017/5/18/15655638/charles-murray -race-iq-sam-harris-science-free-speech.

58. Allen Buchanan et al., *From Chance to Choice: Genetics and Justice* (Cambridge, UK: Cambridge University Press, 2000).

59. There remains frustratingly little consensus about the best language to use to describe patterns of genetic ancestry. I am following convention to describe people with certain patterns of genetic ancestry using the continental adjective "European," but I recognize that this language is imprecise, is likely to have different intuitive meanings for different readers, and risks reifying social categories of race as "pure" biological entities. I return to these issues in more detail in chapter 4. Adam Auton et al., "A Global Reference for Human Genetic Variation," *Nature* 526, no. 7571 (October 2015): 68–74, https://doi.org/10.1038 /nature15393.

Chapter 2: The Genetic Lottery

1. Roberto Tuchman and Isabelle Rapin, "Epilepsy in Autism," *The Lancet Neurology* 1, no. 6 (October 1, 2002): 352–58, https://doi.org/10.1016/S1474 -4422(02)00160-6.

2. Christine A. Olson et al., "The Gut Microbiota Mediates the Anti-Seizure Effects of the Ketogenic Diet," *Cell* 173, no. 7 (June 14, 2018): 1728–41.e13, https://doi.org/10.1016/j.cell.2018.04.027.

3. Emily Perl Kingsley, "Welcome to Holland," *Contact* 136, no. 1 (January 2001): 14, https://doi.org/10.1080/13520806.2001.11758925.

4. Tara Lakes, "I'm Tired of Holland and I Want to Go Home," *Grace for That* (blog), June 10, 2015, https://momlakes.wordpress.com/2015/06/10/im-tired-of-holland-and-i-want-to-go-home/.

5. Raj Rai and Lesley Regan, "Recurrent Miscarriage," *The Lancet* 368, no. 9535 (August 12, 2006): 601–11, https://doi.org/10.1016/S0140-6736(06)69204-0.

6. Emily A. Willoughby et al., "Free Will, Determinism, and Intuitive Judgments About the Heritability of Behavior," *Behavior Genetics* 49, no. 2 (March 2019): 136–53, https://doi.org/10.1007/s10519-018-9931-1.

7. Eric R. Olson, "Why Are Over 250 Million Sperm Cells Released from the Penis during Sex?," Scienceline, June 2, 2008, https://scienceline.org/2008/06/ask-olson-sperm/.

8. Sean B. Carroll, *A Series of Fortunate Events: Chance and the Making of the Planet, Life, and You* (Princeton, NJ: Princeton University Press, 2020).

9. "The American Family Today," Pew Research Center Social & Demographic Trends, December 17, 2015, https://www.pewsocialtrends.org/2015/12/17/1-the-american-family-today/.

10. Lisa Pickoff-White and Ryan Levi, "Are There Really More Dogs Than Children in S.F.?," KQED, May 24, 2018, https://www.kqed.org/news/11669269/are-there-really-more-dogs-than-children-in-s-f.

11. Naomi R. Wray et al., "Complex Trait Prediction from Genome Data: Contrasting EBV in Livestock to PRS in Humans," *Genetics* 211, no. 4 (April 1, 2019): 1131–41, https://doi.org/10.1534/genetics.119.301859.

12. Wray et al.

13. Names have been changed to protect privacy.

14. Francis Galton, *Natural Inheritance* (New York and London: Macmillan, 1894).

15. C. P. Blacker, "The Sterilization Proposals," *The Eugenics Review* 22, no. 4 (January 1931): 239–47.

16. A.W.F. Edwards, "Ronald Aylmer Fisher," in *Time Series and Statistics*, ed. John Eatwell, Murray Milgate, and Peter Newman, first published in *The New Palgrave: A Dictionary of Economics* (London: Palgrave Macmillan UK, 1990), 95–97, https://doi.org/10.1007/978-1-349-20865-4_10.

17. R. A. Fisher, "XV.—The Correlation between Relatives on the Supposition of Mendelian Inheritance," *Earth and Environmental Science Transactions of The Royal Society of Edinburgh* 52, no. 2 (1918): 399–433, https://doi.org/10.1017/S0080456800012163.

18. Ben Cohen, "Shawn Bradley Is Really, Really Tall. But Why?," *Wall Street Journal*, September 18, 2018, https://www.wsj.com/articles/shawn-bradley-genetic-test-height-1537278144.

19. Corinne E. Sexton et al., "Common DNA Variants Accurately Rank an Individual of Extreme Height," *International Journal of Genomics* 2018 (September 4, 2018): 5121540, https://doi.org/10.1155/2018/5121540.

20. "Biologists Checked Out This NBA Player's DNA for Clues to His Immense Height," *MIT Technology Review*, September 1, 2018, https://www .technologyreview.com/s/612014/biologists-checked-out-this-nba-players-dna -for-clues-to-his-immense-height/.

21. Cohen, "Shawn Bradley Is Really, Really Tall. But Why?"

22. Throughout the book, I will use words like "parents," "children," "family," and "siblings" in a narrow sense, to refer to people who are related to each other via the processes of genetic inheritance. This is not to deny the importance of the social relationships that define "family," but instead simply reflects the book's focus on the effects of genetics.

23. "ALDH2 Gene," Genetics Home Reference, accessed July 28, 2020, https:// ghr.nlm.nih.gov/gene/ALDH2.

24. D. Hamer and L. Sirota, "Beware the Chopsticks Gene," *Molecular Psychiatry* 5, no. 1 (January 2000): 11–13, https://www.nature.com/articles/4000662.

25. Simon Haworth et al., "Apparent Latent Structure within the UK Biobank Sample Has Implications for Epidemiological Analysis," *Nature Communications* 10, no. 1 (January 18, 2019): 333, https://doi.org/10.1038/s41467-018-08219-1.

26. Daniel Barth, Nicholas W. Papageorge, and Kevin Thom, "Genetic Endowments and Wealth Inequality," *The Journal of Political Economy* 128, no. 4 (April 2020): 1474–1522, https://doi.org/10.1086/705415.

27. Polygenic indices are more commonly referred to as "polygenic scores." As applied to information about human DNA, however, the word "score" might imply a hierarchy of value. Following the suggestions of my colleagues Patrick Turley and Dan Benjamin, I use the alternative language of "polygenic index" throughout.

28. Daniel W. Belsky et al., "Genetic Analysis of Social-Class Mobility in Five Longitudinal Studies," *Proceedings of the National Academy of Sciences* 115, no. 31 (July 31, 2018): E7275–84, https://doi.org/10.1073/pnas.1801238115.

29. Arthur S. Goldberger, "Heritability," *Economica* 46, no. 184 (1979): 327–47, https://doi.org/10.2307/2553675.

30. George E. P. Box, "Science and Statistics," *Journal of the American Statistical Association* 71, no. 356 (December 1976): 791–99, https://doi.org/10.1080 /01621459.1976.10480949.

Chapter 3: Cookbooks and College

1. "Neurofibromatosis Type 1," Genetics Home Reference, accessed November 7, 2019, https://ghr.nlm.nih.gov/condition/neurofibromatosis-type-1.

2. John Milton, *Lycidas*, accessed November 7, 2019, https://www.poetry foundation.org/poems/44733/lycidas.

3. Cornelius A. Rietveld et al., "GWAS of 126,559 Individuals Identifies Genetic Variants Associated with Educational Attainment," *Science* 340, no. 6139 (June 21, 2013): 1467–71, https://doi.org/10.1126/science.1235488.

4. Avshalom Caspi et al., "Influence of Life Stress on Depression: Moderation by a Polymorphism in the 5-HTT Gene," *Science* 301, no. 5631 (July 18, 2003): 386–89, https://doi.org/10.1126/science.1083968.

5. Richard Border et al., "No Support for Historical Candidate Gene or Candidate Gene-by-Interaction Hypotheses for Major Depression Across Multiple Large Samples," *The American Journal of Psychiatry* 176, no. 5 (May 1, 2019): 376–87, https://doi.org/10.1176/appi.ajp.2018.18070881.

6. Scott Alexander [Siskind], "5-HTTLPR: A Pointed Review," *Slate Star Codex*, May 8, 2019, https://slatestarcodex.com/2019/05/07/5-httlpr-a-pointed-review/.

7. Caspi et al., "Influence of Life Stress on Depression"; Border et al., "No Support for Historical Candidate Gene or Candidate Gene-by-Interaction Hypotheses for Major Depression."

8. Naomi R. Wray et al., "Genome-Wide Association Analyses Identify 44 Risk Variants and Refine the Genetic Architecture of Major Depression," *Nature Genetics* 50, no. 5 (May 2018): 668–81, https://doi.org/10.1038/s41588-018-0090-3.

9. Evan A. Boyle, Yang I. Li, and Jonathan K. Pritchard, "An Expanded View of Complex Traits: From Polygenic to Omnigenic," *Cell* 169, no. 7 (June 15, 2017): 1177–86, https://doi.org/10.1016/j.cell.2017.05.038.

10. James J. Lee et al., "Gene Discovery and Polygenic Prediction from a Genome-Wide Association Study of Educational Attainment in 1.1 Million Individuals," *Nature Genetics* 50, no. 8 (August 2018): 1112–21, https://doi.org/10.1038/s41588-018-0147-3.

11. Rietveld et al., "GWAS of 126,559 Individuals Identifies Genetic Variants Associated with Educational Attainment"; Aysu Okbay et al., "Genome-Wide Association Study Identifies 74 Loci Associated with Educational Attainment," *Nature* 533, no. 7604 (May 2016): 539–42, https://doi.org/10.1038/nature17671; Lee et al.

12. A. G. Allegrini et al., "Genomic Prediction of Cognitive Traits in Childhood and Adolescence," *Molecular Psychiatry* 24, no. 6 (June 2019): 819–27, https://doi.org/10.1038/s41380-019-0394-4.

13. Robert Plomin, *Blueprint: How DNA Makes Us Who We Are* (Cambridge, MA: MIT Press, 2018).

14. David C. Funder and Daniel J. Ozer, "Evaluating Effect Size in Psychological Research: Sense and Nonsense," *Advances in Methods and Practices in Psychological Science* 2, no. 2 (June 1, 2019): 156–68, https://doi.org/10.1177/2515245919847202.

15. Lee et al., "Gene Discovery and Polygenic Prediction from a Genome-Wide Association Study of Educational Attainment in 1.1 Million Individuals."

16. Funder and Ozer, "Evaluating Effect Size in Psychological Research."

17. Matthew J. Salganik et al., "Measuring the Predictability of Life Outcomes with a Scientific Mass Collaboration," *Proceedings of the National Academy of Sciences* 117, no. 15 (April 14, 2020): 8398–8403, https://doi.org/10.1073/pnas.1915006117.

18. Salganik et al.

Chapter 4: Ancestry and Race

1. Aaron Panofsky and Joan Donovan, "Genetic Ancestry Testing among White Nationalists: From Identity Repair to Citizen Science," *Social Studies of Science* 49, no. 5 (October 1, 2019): 653–81, https://doi.org/10.1177/0306312719861434; Jedidiah Carlson and Kelley Harris, "Quantifying and Contextualizing the Impact

of BioRxiv Preprints through Automated Social Media Audience Segmentation," *PLOS Biology* 18, no. 9 (September 22, 2020): e3000860, https://doi.org/10.1371/journal.pbio.3000860.

2. Alex Shoumatoff, "The Mountain of Names," The New Yorker, May 6, 1985, 51ff., https://www.newyorker.com/magazine/1985/05/13/the-mountain-of-names.

3. Quoctrung Bui and Claire Cain Miller, "The Typical American Lives Only 18 Miles From Mom," *The New York Times*, December 23, 2015, https://www.nytimes.com/interactive/2015/12/24/upshot/24up-family.html.

4. Douglas L. T. Rohde, Steve Olson, and Joseph T. Chang, "Modelling the Recent Common Ancestry of All Living Humans," *Nature* 431, no. 7008 (September 30, 2004): 562–66, https://doi.org/10.1038/nature02842; Graham Coop, "Our Vast, Shared Family Tree.," *gcbias* (blog), November 20, 2017, https://gcbias.org/2017/11/20/our-vast-shared-family-tree/.

5. Coop.

6. Dorothy Roberts, *Fatal Invention: How Science, Politics, and Big Business Re-Create Race in the Twenty-First Century* (New York and London: The New Press, 2011).

7. Michael Yudell et al., "Taking Race out of Human Genetics," *Science* 351, no. 6273 (February 5, 2016): 564–65, http://www.ask-force.org/web/Golden-Rice/Yudell-Taking-Race-out-of-human-genetics-2016.pdf.

8. Sam Harris, *Making Sense Podcast* #73, "Forbidden Knowledge," April 22, 2017, https://samharris.org/podcasts/forbidden-knowledge/.

9. Audrey Smedley and Brian D. Smedley, "Race as Biology Is Fiction, Racism as a Social Problem Is Real: Anthropological and Historical Perspectives on the Social Construction of Race," *American Psychologist* 60, no. 1, special issue: Genes, Race, and Psychology in the Genome Era (January 2005): 16–26, https://doi.org/10.1037/0003-066X.60.1.16.

10. Yambazi Banda et al., "Characterizing Race/Ethnicity and Genetic Ancestry for 100,000 Subjects in the Genetic Epidemiology Research on Adult Health and Aging (GERA) Cohort," *Genetics* 200, no. 4 (August 1, 2015): 1285–95, https://doi.org/10.1534/genetics.115.178616.

11. Carl Campbell Brigham, *A Study of American Intelligence* (Princeton, NJ: Princeton University Press, 1922).

12. Noel Ignatiev, *How the Irish Became White* (New York: Routledge, 1995).

13. The 1000 Genomes Project Consortium, "A Global Reference for Human Genetic Variation," *Nature* 526, no. 7571 (October 2015): 68–74, https://doi.org/10.1038/nature15393.

14. United States Census Bureau, "Race: About This Topic," accessed November 7, 2019, https://www.census.gov/topics/population/race/about.html.

15. Banda et al., "Characterizing Race/Ethnicity and Genetic Ancestry for 100,000 Subjects in the Genetic Epidemiology Research on Adult Health and Aging (GERA) Cohort."

16. Alkes L. Price et al., "Principal Components Analysis Corrects for Stratification in Genome-Wide Association Studies," *Nature Genetics* 38, no. 8 (August 2006): 904–9, https://doi.org/10.1038/ng1847.

17. Clare Bycroft et al., "The UK Biobank Resource with Deep Phenotyping and Genomic Data," *Nature* 562, no. 7726 (October 2018): 203–9, https://doi.org /10.1038/s41586-018-0579-z.

18. Yudell et al., "Taking Race out of Human Genetics."

19. Dalton Conley and Jason Fletcher, "What Both the Left and Right Get Wrong About Race," *Nautilus*, June 1, 2017, http://nautil.us/issue/48/chaos/what -both-the-left-and-right-get-wrong-about-race.

20. The 1000 Genomes Project Consortium, "A Global Reference for Human Genetic Variation."

21. Cheryl Stewart and Michael S. Pepper, "Cystic Fibrosis in the African Diaspora," *Annals of the American Thoracic Society* 14, no. 1 (January 2017): 1–7, https://doi.org/10.1513/AnnalsATS.201606-481FR; Giorgio Sirugo, Scott M. Williams, and Sarah A. Tishkoff, "The Missing Diversity in Human Genetic Studies," *Cell* 177, no. 1 (March 21, 2019): 26–31, https://doi.org/10.1016/j.cell.2019.02.048.

22. Nicholas G. Crawford et al., "Loci Associated with Skin Pigmentation Identified in African Populations," *Science* 358, no. 6365 (November 17, 2017), https://doi.org/10.1126/science.aan8433; Sirugo, Williams, and Tishkoff, "The Missing Diversity in Human Genetic Studies."

23. Michael C. Campbell and Sarah A. Tishkoff, "African Genetic Diversity: Implications for Human Demographic History, Modern Human Origins, and Complex Disease Mapping," *Annual Review of Genomics and Human Genetics* 9 (September 22, 2008): 403–33, https://doi.org/10.1146/annurev.genom.9.081307 .164258.

24. L. Duncan et al., "Analysis of Polygenic Risk Score Usage and Performance in Diverse Human Populations," *Nature Communications* 10 (July 25, 2019): 3328, https://doi.org/10.1038/s41467-019-11112-0.

25. James J. Lee et al., "Gene Discovery and Polygenic Prediction from a Genome-Wide Association Study of Educational Attainment in 1.1 Million Individuals," *Nature Genetics* 50, no. 8 (August 2018): 1112–21, https://doi.org/10.1038 /s41588-018-0147-3.

26. Alicia R. Martin et al., "Clinical Use of Current Polygenic Risk Scores May Exacerbate Health Disparities," *Nature Genetics* 51, no. 4 (April 2019): 584–91, https://doi.org/10.1038/s41588-019-0379-x; Duncan et al., "Analysis of Polygenic Risk Score Usage and Performance in Diverse Human Populations."

27. Martin et al., "Clinical Use of Current Polygenic Risk Scores May Exacerbate Health Disparities."

28. W. S. Robinson, "Ecological Correlations and the Behavior of Individuals," *American Sociological Review* 15, no. 3 (June 1950): 351–57.

29. Arthur Jensen, "How Much Can We Boost IQ and Scholastic Achievement?," *Harvard Educational Review* 39, no. 1 (Winter 1969): 1–123, https://doi .org/10.17763/haer.39.1.l3u15956627424k7.

30. . Richard J. Herrnstein and Charles Murray, *The Bell Curve: Intelligence and Class Structure in American Life* (New York: Free Press, 1994).

31. John Novembre and Nicholas H. Barton, "Tread Lightly Interpreting Polygenic Tests of Selection," *Genetics* 208, no. 4 (April 1, 2018): 1351–55, https://doi .org/10.1534/genetics.118.300786.

32. David Reich, "How Genetics Is Changing Our Understanding of 'Race,'" *The New York Times*, March 23, 2018, https://www.nytimes.com/2018/03/23/opinion/sunday/genetics-race.html.

33. Sam Harris, "A Conversation with Kathryn Paige Harden," Making Sense, July 29, 2020, https://samharris.org/subscriber-extras/212-july-29-2020/.

34. Ibram X. Kendi, *How to Be an Antiracist* (New York: One World, 2019).

35. The philosopher Thomas Nagel described how interest in "innate" or "biological" differences between races was tied in people's minds to the question of responsibility: "If one believes that society's responsibility . . . extends only to those disadvantages caused by social injustice, one will assign political importance to the degree, if any, to which racial differences in average I.Q. are genetically influenced." Thomas Nagel, *Mortal Questions* (Cambridge, UK, and New York: Cambridge University Press, 1979).

36. "Paperback Nonfiction Books—Best Sellers," *The New York Times*, July 26, 2020, https://www.nytimes.com/books/best-sellers/2020/07/26/paperback-nonfiction/; Ijeoma Oluo, *So You Want to Talk About Race*, illustrated ed. (Seal Press, 2019); Robin DiAngelo, *White Fragility: Why It's So Hard for White People to Talk About Racism*, foreword by Michael Eric Dyson (Boston: Beacon Press, 2018).

37. Kate Manne, *Down Girl: The Logic of Misogyny* (New York: Oxford University Press, 2017).

38. Theodosius Dobzhansky, "Genetics and Equality: Equality of Opportunity Makes the Genetic Diversity among Men Meaningful," *Science* 137, no. 3524 (July 13, 1962): 112–15, https://doi.org/10.1126/science.137.3524.112.

Chapter 5: A Lottery of Life Chances

1. Amy Mackinnon, "What Actually Happens When a Country Bans Abortion," *Foreign Policy* (blog), May 16, 2019, https://foreignpolicy.com/2019/05/16/what-actually-happens-when-a-country-bans-abortion-romania-alabama/.

2. Vlad Odobescu, "Half a Million Kids Survived Romania's 'Slaughterhouses of Souls.' Now They Want Justice," The World, GlobalPost, PRX (Public Radio Exchange), December 28, 2015, https://www.pri.org/stories/2015-12-28/half-million-kids-survived-romanias-slaughterhouses-souls-now-they-want-justice.

3. Harry F. Harlow, "Love in Infant Monkeys," *Scientific American* 200, no. 6 (June 1959): 68–75.

4. Inge Bretherton, "The Origins of Attachment Theory: John Bowlby and Mary Ainsworth," *Developmental Psychology* 28, no. 5 (September 1992): 759–75, https://doi.org/10.1037/0012-1649.28.5.759.

5. Charles H. Zeanah et al., "Designing Research to Study the Effects of Institutionalization on Brain and Behavioral Development: The Bucharest Early Intervention Project," *Development and Psychopathology* 15, no. 4 (December 2003): 885–907, https://doi.org/10.1017/S0954579403000452.

6. Charles H. Zeanah, Nathan A. Fox, and Charles A. Nelson, "The Bucharest Early Intervention Project: Case Study in the Ethics of Mental Health Research," *The Journal of Nervous and Mental Disease* 200, no. 3 (March 2012): 243–47, https://doi.org/10.1097/NMD.0b013e318247d275; Stephen T. Ziliak and Edward R.

Teather-Posadas, "The Unprincipled Randomization Principle in Economics and Medicine," in *The Oxford Handbook of Professional Economic Ethics, ed.* George F. DeMartino and Deirdre N. McCloskey (New York: Oxford University Press, 2016).

7. Charles A. Nelson et al., "Cognitive Recovery in Socially Deprived Young Children: The Bucharest Early Intervention Project," *Science* 318, no. 5858 (December 21, 2007): 1937–40, https://doi.org/10.1126/science.1143921.

8. David Hume, *An Enquiry concerning Human Understanding,* ed. Peter Millican (New York: Oxford University Press, 2008; orig. pub. 1748).

9. David Lewis, "Causation," *Journal of Philosophy* 70, no. 17 (October 1973): 556–67, https://people.stfx.ca/cbyrne/Byrne/Lewis%20-%20Causation.pdf.

10. John Stuart Mill, "A System of Logic: Ratiocinative and Inductive," in *Collected Works of John Stuart Mill,* vol. 7 (Toronto: University of Toronto Press, 1974), 327, https://oll.libertyfund.org/title/mill-the-collected-works-of-john-stuart-mill -volume-vii-a-system-of-logic-part-i.

11. Donald B. Rubin, "Estimating Causal Effects of Treatments in Randomized and Nonrandomized Studies," *Journal of Educational Psychology* 66, no. 5 (1974): 688–701, https://doi.org/10.1037/h0037350.

12. Paul W. Holland, "Statistics and Causal Inference," *Journal of the American Statistical Association* 81, no. 396 (1986): 945–60, https://doi.org/10.2307 /2289064.

13. More specifically, this method allows you to estimate the *average treatment effect* (ATE). However, the ATE is not the only quantity that researchers might be interested in estimating. For example, they might be specifically interested in heterogeneity in treatment response. For further discussion, see Angus Deaton and Nancy Cartwright, "Understanding and Misunderstanding Randomized Controlled Trials," *Social Science & Medicine* 210, special issue: Randomized Controlled Trials and Evidence-based Policy: A Multidisciplinary Dialogue (August 2018): 2–21, https://doi.org/10.1016/j.socscimed.2017.12.005.

14. Kevin Hartnett, "To Build Truly Intelligent Machines, Teach Them Cause and Effect," *Quanta* Magazine, May 15, 2018, https://www.quantamagazine.org/to -build-truly-intelligent-machines-teach-them-cause-and-effect-20180515/.

15. The evolutionary biologist Richard Dawkins made the point that genetic causes should be defined as difference makers even for relatively simple phenotypes that are intuitively "genetic," such as eye color. He wrote, "The 'effect' of any would-be cause can be given meaning only in terms of a comparison, even if only an implied comparison, with at least one alternative cause. It is strictly incomplete to speak of blue eyes as 'the effect' of a given gene *G1.* If we say such a thing, we really imply the potential existence of at least one alternative allele, call it *G2,* and at least one alternative phenotype, *P2,* in this case, say, brown eyes."

He continues with an example of two genes both related to skin pigmentation: "To be sure, *A,* the gene whose protein product is the black pigment, is necessary in order for an individual to be black. . . . But I shall not call *A* a gene for blackness unless some of the variation in the population is due to lack of *A.* . . . The point that is relevant here is that both *A* and *B* are potentially entitled to be called genes for blackness, *depending on the alternatives that exist in the population* (emphasis added). The fact that the causal chain linking *A* to the production of the black pigment molecule is short, while that for *B* is long and tortuous, is irrelevant."

Finally, Dawkins pointed out that natural selection is concerned with differences: some versions of genes become more common than others because those versions cause differences in fitness. Evolution requires a comparison.

Failure to appreciate the fact that genetic causes, like all other causes, are difference makers that imply a comparison to some alternative is one major flaw in the reasoning of a still widely cited essay by the philosopher Ned Block. He wrote (emphasis added), "Genetic determination is a matter of *what causes a characteristic*: number of toes is genetically determined because our genes cause us to have five toes. Heritability, by contrast, is a matter of what *causes differences in a characteristic*: heritability of number of toes is a matter of the extent to which genetic differences cause variation in number of toes (that some cats have five toes, and some have six)." Block's error should be readily apparent. What causes a characteristic *is*, by definition, what causes differences in a characteristic. To say that a gene *G1* causes us to have five toes is to imply the existence of an alternative allele and an alternative phenotype—having a gene other than *G1* would cause you to have a different number of toes.

In fact, the fact that genes are difference makers can be empirically illustrated using Block's example of having five toes. Two of the genes that determine toe number are *EVC1* and *EVC2*. Rare mutations in these genes cause polydactyly (extra fingers and toes), as well as short ribs, teeth abnormalities, and cardiac defects, a syndrome known as Ellis–van Creveld syndrome. The *EVC* genes code for a protein that is found on the little hairlike projections that surround each cell; the protein helps cells communicate with each other so that they can arrange themselves into the right shapes. The *EVC1* and *EVC2* genes were discovered by studying nine Amish families in which some family members were born with extra fingers and toes. Scientists in this study focused on the exact question that Block, wrongly, identified as distinct from the question of genetic causation: they asked, What genes are associated with a difference in whether or not you have five fingers and five toes? Those who inherited two copies of a mutated form of the *EVC1* or *EVC2* genes had extra toes; those who didn't had five toes.

Richard Dawkins, *The Extended Phenotype: The Long Reach of the Gene*, rev. ed. (Oxford and New York: Oxford University Press, 1999); Ned Block, "How Heritability Misleads about Race," *The Boston Review* 20, no. 6 (January 1996): 30–35; Victor A. McKusick, "Ellis-van Creveld Syndrome and the Amish," *Nature Genetics* 24, no. 3 (March 2000): 203–4, https://doi.org/10.1038/73389.

16. John March et al., "Fluoxetine, Cognitive-Behavioral Therapy, and Their Combination for Adolescents with Depression: Treatment for Adolescents With Depression Study (TADS) Randomized Controlled Trial," *JAMA* 292, no. 7 (August 1, 2004): 807–20, https://doi.org/10.1001/jama.292.7.807.

17. Robert Ross et al., "Reduction in Obesity and Related Comorbid Conditions after Diet-Induced Weight Loss or Exercise-Induced Weight Loss in Men: A Randomized Controlled Trial," *Annals of Internal Medicine* 133, no. 2 (July 18, 2000): 92–103, https://doi.org/10.7326/0003-4819-133-2-200007180-00008.

18. MRC Vitamin Research Study Group1, "Prevention of Neural Tube Defects: Results of the Medical Research Council Vitamin Study," *The Lancet* 338, no. 8760 (July 20, 1991): 131–37, https://doi.org/10.1016/0140-6736(91)90133-A.

19. Urie Bronfenbrenner and Pamela L. Morris, "The Bioecological Model of Human Development," in *Handbook of Child Psychology*, vol. 1, *Theoretical Models of Human Development,* ed. Richard M. Lerner and William Damon, 6th ed. (Hoboken, NJ: John Wiley and Sons, 2007), https://onlinelibrary.wiley.com/doi/abs/10.1002/9780470147658.chpsy0114.

20. Pamela Herd et al., "Genes, Gender Inequality, and Educational Attainment," *American Sociological Review* 84, no. 6 (December 1, 2019): 1069–98, https://doi.org/10.1177/0003122419886550.

21. Richard C. Lewontin, "The Analysis of Variance and the Analysis of Causes," *International Journal of Epidemiology* 35, no. 3 (June 2006): 520–25, https://doi.org/10.1093/ije/dyl062.

22. Clifford Geertz, "Thick Description: Toward an Interpretive Theory of Culture," in *The Interpretation of Culture* (New York: Basic Books, 1973), https://philpapers.org/archive/geettd.pdf. I am grateful to Benjamin Domingue for pointing out the similarities between my language here and Geertz's distinction between "thin" and "thick" description of behavior, e.g., "rapidly contracting his right eyelids" versus "practicing a burlesque of a friend faking a wink to deceive an innocent into thinking a conspiracy is in motion."

Chapter 6: Random Assignment by Nature

1. Peter M. Visscher et al., "Assumption-Free Estimation of Heritability from Genome-Wide Identity-by-Descent Sharing between Full Siblings," *PLOS Genetics* 2, no. 3 (March 24, 2006): e41, https://doi.org/10.1371/journal.pgen.0020041.

2. Nancy L. Segal, *Born Together—Reared Apart: The Landmark Minnesota Twin Study*, illustrated edition (Cambridge, MA: Harvard University Press, 2012).

3. *Three Identical Strangers* (2018), IMDb, accessed February 9, 2021, https://www.imdb.com/title/tt7664504/.

4. Tinca J. C. Polderman et al., "Meta-Analysis of the Heritability of Human Traits Based on Fifty Years of Twin Studies," *Nature Genetics* 47, no. 7 (July 2015): 702–9, https://doi.org/10.1038/ng.3285.

5. Sophie von Stumm, Benedikt Hell, and Tomas Chamorro-Premuzic, "The Hungry Mind: Intellectual Curiosity Is the Third Pillar of Academic Performance," *Perspectives on Psychological Science* 6, no. 6 (November 1, 2011): 574–88, https://doi.org/10.1177/1745691611421204.

6. Richard C. Lewontin, "The Analysis of Variance and the Analysis of Causes," *International Journal of Epidemiology* 35, no. 3 (June 2006): 520–25, https://doi.org/10.1093/ije/dyl062

7. Richard M. Lerner, "Another Nine-Inch Nail for Behavioral Genetics!," *Human Development* 49, no. 6 (2007): 336–42, https://doi.org/DOI:10.1159/000096532.

8. Charles F. Manski, "Genes, Eyeglasses, and Social Policy," *Journal of Economic Perspectives* 25, no. 4 (Fall 2011): 83–94, https://doi.org/10.1257/jep.25.4.83.

9. Another objection: it doesn't matter that these traits are heritable because everything is heritable. That is, everything you can measure about a person that differs within a population shows some evidence of heritable variation. This extends to even silly traits, like how much TV you watch or how much Marmite you like to eat. Silly examples are useful in pushing back against the intuition, which I discussed in the last chapter, that genetic *causation* implies something like a biodeterminist *mechanism.* We are not going to understand Marmite-liking and TV-watching "at the level of the genome." But we don't care about the heritability of Marmite-liking, not because heritability is a useless and "metaphorical" statistic, but because we don't care whether people like Marmite or not. We do care, however, whether or not people graduate from college. The scientific and philosophical importance of heritability statistics is derived from the scientific and philosophical importance of the phenotype. Eric Turkheimer, "Three Laws of Behavior Genetics and What They Mean," *Current Directions in Psychological Science* 9, no. 5 (October 1, 2000), 160–64, https://journals.sagepub.com/doi/10.1111/1467-8721.00084.

10. The connection between heritability and genetic causation can be further clarified by considering how heritability coefficients are used in agricultural selection programs. The so-called "breeder's equation" is given as: $R = h^2 \times S$, where h^2 is the heritability coefficient in a population, R is the response to selection, defined as the change in the mean phenotype between generations, and S is how different the parents who are selected for breeding are from the mean in the population.

In the United States in 2019, the mean height for men is 5′9″ (176 cm). Imagine that a dystopian dictatorship ruled that only men who were taller than a certain threshold were allowed to father children. As a result, the average height among fathers selected for breeding was 6′0″. The difference between parents selected for breeding and the mean in the population is, in this instance, 3 inches. Assuming mothers were subject to selection of similar magnitude, how much taller will the next generation of male children be, on average, than they would have been in the absence of selection on the parents, assuming that everything about the environment is kept exactly the same? The heritability of height, according to the Visscher study that I described at the beginning of this chapter, was estimated to be 0.80. That's not 1.0—the next generation of sons won't also be 3 inches taller, on average. But a high heritability means that the offspring of the selectively bred parents will, in fact, be substantially taller—just over 2 inches on average. A shift in the mean of the population has implications for how frequently "extreme" values are observed. In a population with a mean height of 69 inches, about 1 percent of men are taller than 6′6″. Shift the mean height up 2 inches to 71 inches, and now about 4 percent of men are that tall.

Because it determines response to selection, the causal relevance of heritability can be further understood within the framework of the *manipulationist theory of causation.* Related to the theories of causation as counterfactual dependence that I described in the previous chapter, the manipulationist theory is not centered on the question, "What would have happened to Y if X had not happened?," but is rather centered on the question, "What would happen to Y if you changed X?"

The philosopher Jim Woodward describes this more precisely in *Making Things Happen*: "The claim that X causes Y means that for at least some individuals, there is a possible manipulation of some value of X that they possess, which, given other appropriate conditions (perhaps including manipulations that fix other variables distinct from X at certain values), will change the value of Y or the probability distribution of Y for those individuals" (p. 40).

Selection experiments are an interesting twist on this requirement. The claim that genes (X) cause the phenotype (Y) means that for at least some individuals, there is a possible manipulation of some value of X that they possess. In the case of selection, this manipulation is to restrict the range of genotypes allowed to reproduce. Given other appropriate conditions, including fixing other variables distinct from X (i.e., environmental conditions) at certain values, this will change the probability distribution of Y for those individuals' offspring.

If selection experiments demonstrate the causal power of genes for the phenotype, and heritability determines the response to selection, it is impossible to conclude that heritability is somehow irrelevant to causation. As Peter Visscher described in another paper, "Heritability is a fundamental parameter in genetics . . . it is key to selection in evolutionary biology and agriculture, and to the prediction of disease risk in medicine."

James Woodward, *Making Things Happen: A Theory of Causal Explanation*, Oxford Studies in Philosophy of Science (Oxford: Oxford University Press, 2003); Peter M. Visscher, William G. Hill, and Naomi R. Wray, "Heritability in the Genomics Era—Concepts and Misconceptions," *Nature Reviews Genetics* 9, no. 4 (April 2008): 255–66, https://doi.org/10.1038/nrg2322.

11. The equal environments assumption has been the subject of much scrutiny, and newer studies taking advantage of measured DNA have largely found support for it. One notable study took advantage of the fact that parents, pediatricians, and even twins themselves frequently misclassify zygosity—they think they are identical when they are actually fraternal, or vice versa. One study of about 300 Dutch twins found that parents were wrong about their children's zygosity 19 percent of the time. I find a similar thing in the twin study that I run in Texas: college students who have met a set of twins once are better than the twins' parents at guessing whether DNA results will show the twins to be identical or fraternal. The sociologist Dalton Conley and his colleagues leveraged this parental bias in order to test the equal environments assumption, reasoning that if identical twins are more similar than fraternal twins because their parents treat them more similarly (a violation of the equal environments assumption), then twin pairs that are *really* fraternal, but who have been misclassified as identical, will be more similar to one another than are twin pairs who have been correctly classified as fraternal. This is, in fact, what Conley was hoping to find. To a sociologist trained to view the results of behavior genetics with fear and loathing, the design seemed like a clever way to undermine the steadily mounting evidence that genes mattered for understanding social inequality. But that's exactly what he *didn't* find! Instead, the study found that twins' phenotypic similarity (i.e., how similar twins are for their outcomes) tracked their actual genetic relationship, not what their parents thought their zygosity was—evidence in support of the equal environments assumption.

Dalton Conley et al., "Heritability and the Equal Environments Assumption: Evidence from Multiple Samples of Misclassified Twins," *Behavior Genetics* 43, no. 5 (September 2013): 415–26, https://doi.org/10.1007/s10519-013-9602-1.

12. James J. Lee et al., "Gene Discovery and Polygenic Prediction from a Genome-Wide Association Study of Educational Attainment in 1.1 Million Individuals," *Nature Genetics* 50, no. 8 (August 2018): 1112–21, https://doi.org/10.1038/s41588-018-0147-3.

13. Matthew J. Salganik et al., "Measuring the Predictability of Life Outcomes with a Scientific Mass Collaboration," *Proceedings of the National Academy of Sciences* 117, no. 15 (April 14, 2020): 8398–8403, https://doi.org/10.1073/pnas.1915006117.

14. Amelia R. Branigan, Kenneth J. McCallum, and Jeremy Freese, "Variation in the Heritability of Educational Attainment: An International Meta-Analysis," *Social Forces* 92, no. 1 (September 2013): 109–40.

15. Alexander I. Young, "Solving the Missing Heritability Problem," *PLOS Genetics* 15, no. 6 (June 24, 2019): e1008222, https://doi.org/10.1371/journal.pgen.1008222.

16. Young.

17. Alexander I. Young et al., "Relatedness Disequilibrium Regression Estimates Heritability without Environmental Bias," *Nature Genetics* 50, no. 9 (September 2018): 1304–10, https://doi.org/10.1038/s41588-018-0178-9.

18. Lee et al., "Gene Discovery and Polygenic Prediction from a Genome-Wide Association Study of Educational Attainment in 1.1 Million Individuals."

19. Saskia Selzam et al., "Comparing Within- and Between-Family Polygenic Score Prediction," *The American Journal of Human Genetics* 105, no. 2 (August 1, 2019): 351–63, https://doi.org/10.1016/j.ajhg.2019.06.006.

20. Daniel W. Belsky et al., "Genetic Analysis of Social-Class Mobility in Five Longitudinal Studies," *Proceedings of the National Academy of Sciences* 115, no. 31 (July 31, 2018): E7275–84, https://doi.org/10.1073/pnas.1801238115.

21. Rosa Cheesman et al., "Comparison of Adopted and Nonadopted Individuals Reveals Gene–Environment Interplay for Education in the UK Biobank," *Psychological Science* 31, no. 5 (May 1, 2020): 582–91, https://doi.org/10.1177/0956797620904450.

22. Augustine Kong et al., "The Nature of Nurture: Effects of Parental Genotypes," *Science* 359, no. 6374 (January 26, 2018): 424–28, https://doi.org/10.1126/science.aan6877.

23. Theodosius Dobzhansky, "Genetics and Equality: Equality of Opportunity Makes the Genetic Diversity among Men Meaningful," *Science* 137, no. 3524 (July 13, 1962): 112–15, https://doi.org/10.1126/science.137.3524.112.

Chapter 7: The Mystery of How

1. Christopher Jencks et al., *Inequality: A Reassessment of the Effect of Family and Schooling in America* (New York: Basic Books, 1972).

2. Complicated human behaviors are not the only phenotypes that are connected to genotypes via long causal chains. As the evolutionary biologist Richard

Dawkins argued, "What on earth [is] any genetic trait . . . morphological, physiological, or behavioural, if not a 'byproduct' of something more fundamental? If we think the matter through we find that all genetic effects are 'byproducts' except protein molecules." Similarly, it is now becoming clear that even apparently simple environmental interventions can also depend on long causal chains involving complex social processes, such as peer norms and teacher effects, in order to be effective. Richard Dawkins, *The Extended Phenotype: The Long Reach of the Gene*, rev. ed. (Oxford and New York: Oxford University Press, 1999)

3. Paul Oppenheim and Hilary Putnam, "Unity of Science as a Working Hypothesis," in *Concepts, Theories, and the Mind-Body Problem*, Minnesota Studies in the Philosophy of Science, vol. 2 (Minneapolis: University of Minnesota Press, 1958), 3–36, http://conservancy.umn.edu/handle/11299/184622.

4. Carl F. Craver and Lindley Darden, *In Search of Mechanisms: Discoveries across the Life Sciences* (Chicago: University of Chicago Press, 2013).

5. Francis Galton, *Hereditary Genius: An Inquiry into Its Laws and Consequences* (London and New York: Macmillan, 1892).

6. Charles Murray, *Human Diversity: The Biology of Gender, Race, and Class* (New York: Twelve, 2020).

7. Kate Manne, *Down Girl: The Logic of Misogyny* (New York: Oxford University Press, 2017).

8. Theodosius Dobzhansky, "Genetics and Equality: Equality of Opportunity Makes the Genetic Diversity among Men Meaningful," *Science* 137, no. 3524 (July 13, 1962): 112–15, https://doi.org/10.1126/science.137.3524.112.

9. It's important to remember that the problem of unknown mechanisms, which perhaps operate through unintuitive mediators, is not a problem specific to genetic causes. In fact, this problem can attend *any* causal inference made from a randomized controlled trial (RCT). In their review of the strengths and weaknesses of RCTs, the Nobel prize-winning economist Angus Deaton and the philosopher of science Nancy Cartwright argued that "a great deal of other work—empirical, theoretical, and conceptual—needs to be done to make the results of an RCT serviceable." You might know that intervening in this one way under this one set of controlled conditions has this average treatment effect, but what are the boundary conditions? What is the chain of causal events between intervention and eventual outcome? How do people differ in their response to the intervention? So, too, is it insufficient merely to test the average treatment effect of a set of genetic variants on an outcome using nature's randomization. There is empirical, theoretical, and conceptual work to be done to make the results of that causal inference scientifically and practically useful. Deaton and Cartwright, "Understanding and Misunderstanding Randomized Controlled Trials," *Social Science & Medicine* 210, special issue: Randomized Controlled Trials and Evidence-based Policy: A Multidisciplinary Dialogue (August 2018): 2–21, https://doi.org/10.1016/j.socscimed.2017.12.005.

10. James J. Lee et al., "Gene Discovery and Polygenic Prediction from a Genome-Wide Association Study of Educational Attainment in 1.1 Million Individuals," *Nature Genetics* 50, no. 8 (August 2018): 1112–21, https://doi.org/10.1038/s41588-018-0147-3.

11. Elliot M. Tucker-Drob et al., "Emergence of a Gene × Socioeconomic Status Interaction on Infant Mental Ability Between 10 Months and 2 Years," *Psychological Science* 22, no. 1 (January 2011): 125–33, https://doi.org/10.1177/0956797610392926.

12. Daniel W. Belsky et al., "Genetic Analysis of Social-Class Mobility in Five Longitudinal Studies," *Proceedings of the National Academy of Sciences* 115, no. 31 (July 31, 2018): E7275–84, https://doi.org/10.1073/pnas.1801238115; Daniel W. Belsky and K. Paige Harden, "Phenotypic Annotation: Using Polygenic Scores to Translate Discoveries from Genome-Wide Association Studies from the Top Down," *Current Directions in Psychological Science* 28, no. 1 (February 1, 2019): 82–90, https://doi.org/10.1177/0963721418807729; J. Wertz et al., "Genetics and Crime: Integrating New Genomic Discoveries Into Psychological Research About Antisocial Behavior," *Psychological Science* 29, no. 5 (May 1, 2018): 791–803, https://doi.org/10.1177/0956797617744542; Daniel W. Belsky et al., "The Genetics of Success: How Single-Nucleotide Polymorphisms Associated with Educational Attainment Relate to Life Course Development," *Psychological Science* 27, no. 7 (July 1, 2016): 957–72; Emily Smith-Woolley et al., "Differences in Exam Performance between Pupils Attending Selective and Non-Selective Schools Mirror the Genetic Differences between Them," *Npj Science of Learning* 3 (March 2018): 3, https://www.nature.com/articles/s41539-018-0019-8; Eveline L. de Zeeuw et al., "Polygenic Scores Associated with Educational Attainment in Adults Predict Educational Achievement and ADHD Symptoms in Children," *American Journal of Medical Genetics Part B: Neuropsychiatric Genetics* 165B, no. 6 (September 2014), 510–20, https://onlinelibrary.wiley.com/doi/full/10.1002/ajmg.b.32254; Robert Plomin and Sophie von Stumm, "The New Genetics of Intelligence," *Nature Reviews Genetics* 19, no. 3 (March 2018): 148–59, https://doi.org/10.1038/nrg.2017.104; Andrea G. Allegrini et al., "Genomic Prediction of Cognitive Traits in Childhood and Adolescence," *Molecular Psychiatry* 24, no. 6 (June 2019): 819–27, https://www.nature.com/articles/s41380-019-0394-4.

13. Laura E. Engelhardt et al., "Genes Unite Executive Functions in Childhood," *Psychological Science* 26, no. 8 (August 1, 2015): 1151–63, https://doi.org/10.1177/0956797615577209.

14. One common criticism of twin studies is that they might underestimate the extent to which environmental factors shared by kids in the same home contribute to variation in their life outcomes, because the studies don't include sufficiently many families from disadvantaged backgrounds. Remember that heritability is a proportion, and the more environmental variation there is in the sample, the bigger the denominator and the smaller the heritability. In the case of the Texas Twin Project, however, our sample *does* represent a broad range of environmental adversity. One-third of our participating families have received some sort of public assistance (like SNAP, i.e., assistance buying food) since the kids were born. We also calculated the Gini index—a measure of income inequality—of our sample. It was 0.35, compared to 0.39 in the United States as a whole, indicating that we are doing a reasonable job, particularly given the geographical restriction of our sample, of capturing the broader pattern of income inequality that characterizes American society.

The composition of our sample is important, because it means that we don't see the very high heritability of EF just because we've only sampled children who all come from similarly affluent backgrounds. What's more, an independent lab in Colorado, run by the psychologist Naomi Friedman, found the *exact* same result of perfect heritability with a totally different sample of twins who were older at the time they were tested. Naomi P. Friedman et al., "Individual Differences in Executive Functions Are Almost Entirely Genetic in Origin," *Journal of Experimental Psychology: General* 137, no. 2 (May 2008): 201–25, https://doi.org/10.1037/0096-3445.137.2.201.

15. Elliot M. Tucker-Drob and Daniel A. Briley, "Continuity of Genetic and Environmental Influences on Cognition across the Life Span: A Meta-Analysis of Longitudinal Twin and Adoption Studies," *Psychological Bulletin* 140, no. 4 (July 2014): 949–79, https://doi.org/10.1037/a0035893.

16. Fyodor Dostoyevsky, *Crime and Punishment*, translated by Richard Pevear and Larissa Volokhonsky (New York: Alfred A. Knopf, 1991).

17. Paul Tough, *How Children Succeed: Grit, Curiosity, and the Hidden Power of Character* (Houghton Mifflin Harcourt, 2012), https://www.amazon.com/How-Children-Succeed-Curiosity-Character/dp/0544104404.

18. James J. Heckman, "Skill Formation and the Economics of Investing in Disadvantaged Children," *Science* 312, no. 5782 (June 30, 2006): 1900–1902, https://doi.org/10.1126/science.1128898.

19. Carol Dweck, *The Power of Believing That You Can Improve*, TEDx Norrkoping, November 2014, https://www.ted.com/talks/carol_dweck_the_power_of_believing_that_you_can_improve.

20. Tough, *How Children Succeed*.

21. Jonah Lehrer, "Which Traits Predict Success? (The Importance of Grit)," *Wired*, March 14, 2011, https://www.wired.com/2011/03/what-is-success-true-grit/.

22. Belsky et al., "Genetic Analysis of Social-Class Mobility in Five Longitudinal Studies"; Belsky et al., "The Genetics of Success: How SNPs Associated with Educational Attainment Relate to Life Course Development"; Wertz et al., "Genetics and Crime"; Smith-Woolley et al., "Differences in Exam Performance between Pupils Attending Selective and Non-Selective Schools Mirror the Genetic Differences between Them"; de Zeeuw et al., "Polygenic Scores Associated with Educational Attainment in Adults Predict Educational Achievement and ADHD Symptoms in Children"; Plomin and Stumm, "The New Genetics of Intelligence"; Allegrini et al., "Genomic Prediction of Cognitive Traits in Childhood and Adolescence."

23. Perline Demange et al., "Investigating the Genetic Architecture of Noncognitive Skills Using GWAS-by-Subtraction," *Nature Genetics* 53 (January 7, 2021): 35–44, https://doi.org/10.1038/s41588-020-00754-2.

24. Perline Demange et al., "Genetic Associations between Non-Cognitive Skills and Educational Outcomes: The Role of Parental Environment," BGA 2020, Behavior Genetics Association 50th annual meeting, online, June 25–26, 2020, http://bga.org/wp-content/uploads/2020/06/Cheesman_Abstract_BGA2020.pdf.

25. Brendan Bulik-Sullivan et al., "An Atlas of Genetic Correlations across Human Diseases and Traits," *Nature Genetics* 47, no. 11 (November 2015): 1236–41, https://doi.org/10.1038/ng.3406.

26. Demange et al., "Investigating the Genetic Architecture of Non-Cognitive Skills Using GWAS-by-Subtraction."

27. Tucker-Drob and Briley, "Continuity of Genetic and Environmental Influences on Cognition across the Life Span."

28. Elliot M. Tucker-Drob, Daniel A. Briley, and K. Paige Harden, "Genetic and Environmental Influences on Cognition Across Development and Context," *Current Directions in Psychological Science* 22, no. 5 (October 1, 2013): 349–55, https://doi.org/10.1177/0963721413485087.

29. Elliot M. Tucker-Drob and K. Paige Harden, "Early Childhood Cognitive Development and Parental Cognitive Stimulation: Evidence for Reciprocal Gene–Environment Transactions," *Developmental Science* 15, no. 2 (March 2012): 250–59, https://doi.org/10.1111/j.1467-7687.2011.01121.x.

30. Jasmin Wertz et al., "Genetics of Nurture: A Test of the Hypothesis That Parents' Genetics Predict Their Observed Caregiving," *Developmental Psychology* 55, no. 7 (2019): 1461–72, https://doi.org/10.1037/dev0000709.

31. K. Paige Harden et al., "Genetic Associations with Mathematics Tracking and Persistence in Secondary School," *Npj Science of Learning* 5 (February 5, 2020): 1, https://doi.org/10.1038/s41539-020-0060-2.

32. David Lee Stevenson and Kathryn S. Schiller, "State Education Policies and Changing School Practices: Evidence from the National Longitudinal Study of Schools, 1980–1993," *American Journal of Education* 107, no. 4 (August 1999): 261–88.

Chapter 8: Alternative Possible Worlds

1. Arthur Jensen, "How Much Can We Boost IQ and Scholastic Achievement?," *Harvard Educational Review* 39, no. 1 (Winter 1969): 1–123, https://doi.org/10.17763/haer.39.1.l3u15956627424k7.

2. Charles Murray, *Human Diversity: The Biology of Gender, Race, and Class* (New York: Twelve, 2020).

3. Arthur S. Goldberger, "Heritability," *Economica* 46, no. 184 (1979): 327–47, https://doi.org/10.2307/2553675.

4. Heritability does not have clear implications for whether environmentally induced change is possible for a phenotype, but it might have implications for whether those environmentally induced changes persist across generations. Returning to Goldberger's example of eyeglasses, one's own vision can be corrected by eyeglasses, but that improvement in vision will not persist to your children if they are not also given access to eyeglasses. As Conley and Fletcher put it, "Any interventions that prevent or fix [an adverse outcome like poor eyesight] are unlikely to yield dynastic payoffs in the next generation, because the risk inherent in the germ line [i.e., genetically transmitted from parents to offspring] has not

been altered. . . . We will have to keep applying those solutions for each generation if we wanted the beneficial effects to persist." Dalton Conley and Jason Fletcher, *The Genome Factor: What the Social Genomics Revolution Reveals about Ourselves, Our History, and the Future* (Princeton, NJ: Princeton University Press, 2017).

5. Theodosius Dobzhansky, "Genetics and Equality: Equality of Opportunity Makes the Genetic Diversity among Men Meaningful," *Science* 137, no. 3524 (July 13, 1962): 112–15, https://doi.org/10.1126/science.137.3524.112.

6. Stephanie Welch, *A Dangerous Idea: Eugenics, Genetics and the American Dream*, documentary (Paragon Media), accessed November 13, 2019, http://adangerousideafilm.com/.

7. Mikk Titma, Nancy Brandon Tuma, and Kadi Roosma, "Education as a Factor in Intergenerational Mobility in Soviet Society," *European Sociological Review* 19, no. 3 (July 1, 2003): 281–97, https://doi.org/10.1093/esr/19.3.281.

8. OECD, *Equity and Quality in Education: Supporting Disadvantaged Students and Schools* (Paris: OECD Publishing, 2012).

9. Pamela Herd et al., "Genes, Gender Inequality, and Educational Attainment," *American Sociological Review* 84, no. 6 (December 1, 2019): 1069–98, https://doi.org/10.1177/0003122419886550.

10. A. C. Heath et al., "Education Policy and the Heritability of Educational Attainment," *Nature* 314, no. 6013 (April 25, 1985): 734–36, https://doi.org/10.1038/314734a0.

11. Per Engzell and Felix C. Tropf, "Heritability of Education Rises with Intergenerational Mobility," *Proceedings of the National Academy of Sciences* 116, no. 51 (November 29, 2019): 25386–88, https://doi.org/10.1073/pnas.1912998116; Wendy Johnson et al., "Family Background Buys an Education in Minnesota but Not in Sweden," *Psychological Science* 21, no. 9 (September 1, 2010): 1266–73, https://doi.org/10.1177/0956797610379233.

12. Elliot M. Tucker-Drob and Timothy C. Bates, "Large Cross-National Differences in Gene × Socioeconomic Status Interaction on Intelligence," *Psychological Science* 27, no. 2 (February 1, 2016): 138–49, https://doi.org/10.1177/0956797615612727.

13. Ned Block, "How Heritability Misleads about Race," *The Boston Review* 20, no. 6 (January 1996): 30–35.

14. Block.

15. Stephen J. Ceci and Paul B. Papierno, "The Rhetoric and Reality of Gap Closing: When the 'Have-Nots' Gain but the 'Haves' Gain Even More," *American Psychologist* 60, no. 2 (2005): 149–60, https://doi.org/10.1037/0003-066X.60.2.149.

16. Richard J. Herrnstein, *I.Q. in the Meritocracy* (Boston: Little, Brown, 1973).

17. Conley and Fletcher, *The Genome Factor*.

18. Conley and Fletcher.

19. Hiu Man Grisch-Chan et al., "State-of-the-Art 2019 on Gene Therapy for Phenylketonuria," *Human Gene Therapy* 30, no. 10 (October 2019): 1274–83, https://doi.org/10.1089/hum.2019.111.

20. Evan A. Boyle, Yang I. Li, and Jonathan K. Pritchard, "An Expanded View of Complex Traits: From Polygenic to Omnigenic," *Cell* 169, no. 7 (June 2017): 1177–86, https://doi.org/10.1016/j.cell.2017.05.038.

21. V. Bansal et al., "Genome-Wide Association Study Results for Educational Attainment Aid in Identifying Genetic Heterogeneity of Schizophrenia," *Nature Communications* 9, no. 1 (August 6, 2018): 3078, http://dx.doi.org/10.1038/s41467 -018-05510-z; Demange et al., "Investigating the Genetic Architecture of Noncognitive Skills Using GWAS-by-Subtraction."

22. Richard Haier, "No Voice at VOX: Sense and Nonsense about Discussing IQ and Race," *Quillette*, June 11, 2017, https://quillette.com/2017/06/11/no-voice -vox-sense-nonsense-discussing-iq-race/; Ann Brown, "John McWhorter: Racial Equality May Mean Genetic Editing To Close Racial IQ Gap," The Moguldom Nation, February 9, 2021, https://moguldom.com/335699/john-mcwhorter-racial -equality-may-mean-genetic-editing-to-close-racial-iq-gap/.

23. Leon J. Kamin, "Commentary," in Sandra Scarr, *Race, Social Class, and Individual Differences in IQ* (Hillsdale, NJ: Lawrence Erlbaum Associates, 1981), 482.

24. John Rawls, *A Theory of Justice*, rev. ed. (Cambridge, MA: Harvard University Press, 1999).

25. OECD, *Equity in Education: Breaking Down Barriers to Social Mobility*, PISA (Paris: OECD Publishing, 2018), https://doi.org/10.1787/9789264073234 -en.

26. H. Moriah Sokolowski and Daniel Ansari, "Understanding the Effects of Education through the Lens of Biology," *Npj Science of Learning* 3 (October 1, 2018): 17, https://doi.org/10.1038/s41539-018-0032-y; Carina Omoeva, "Mainstreaming Equity in Education," issues paper, FHI 360 Education Equity Research Initiative, September 2017, 26, http://www.educationequity2030.org/resources -2/2017/10/27/mainstreaming-equity-in-education.

27. Richard Arneson, "Four Conceptions of Equal Opportunity," *The Economic Journal* 128, no. 612 (July 1, 2018): F152–73, https://doi.org/10.1111/ecoj.12531.

28. Thomas Nagel, *Mortal Questions* (Cambridge, UK, New York: Cambridge University Press, 1979)

29. Fredrik deBoer, *The Cult of Smart: How Our Broken Education System Perpetuates Social Injustice* (New York: All Points Books, 2020).

30. Silvia H. Barcellos, Leandro S. Carvalho, and Patrick Turley, "Education Can Reduce Health Differences Related to Genetic Risk of Obesity," *Proceedings of the National Academy of Sciences* 115, no. 42 (October 16, 2018): E9765–72, https://doi.org/10.1073/pnas.1802909115.

31. Sally I-Chun Kuo et al., "The Family Check-up Intervention Moderates Polygenic Influences on Long-Term Alcohol Outcomes: Results from a Randomized Intervention Trial," *Prevention Science* 20, no. 7 (October 2019): 975–85, https:// doi.org/10.1007/s11121-019-01024-2.

32. Jason M. Fletcher, "Why Have Tobacco Control Policies Stalled? Using Genetic Moderation to Examine Policy Impacts," *PLOS ONE* 7, no. 12 (December 5, 2012): e50576, https://doi.org/10.1371/journal.pone.0050576.

33. Jason D. Boardman et al., "Population Composition, Public Policy, and the Genetics of Smoking," *Demography* 48, no. 4 (November 2011): 1517–33, https://doi.org/10.1007/s13524-011-0057-9; Benjamin W. Domingue et al., "Cohort Effects in the Genetic Influence on Smoking," *Behavior Genetics* 46, no. 1 (January 2016): 31–42, https://doi.org/10.1007/s10519-015-9731-9.

34. Ceci and Papierno, "The Rhetoric and Reality of Gap Closing."

35. Harris Cooper et al., "Making the Most of Summer School: A Meta-Analytic and Narrative Review," *Monographs of the Society for Research in Child Development* 65, no. 1 (February 2000): i–127; Thomas D. Cook et al., *Sesame Street Revisited* (New York: Russell Sage Foundation, 1975).

36. Anthony J. F. Griffiths et al., "Norm of Reaction and Phenotypic Distribution," in *An Introduction to Genetic Analysis, 7th ed.*, ed. Anthony J. F. Griffiths et al. (New York: W. H. Freeman, 2000), http://www.ncbi.nlm.nih.gov/books/NBK22080/.

37. Most studies of gene × intervention effects or gene × environment effects have used poor measures of genotype (e.g., examining the effects of a single genetic variant) or have used measures of environmental context that are themselves correlated with people's genetic differences. The relatively few well-done studies, in contrast, have good measures of genotype (such as a polygenic index created from a highly powered GWAS) and examine environments using quasi-experimental designs that allow for better causal inference about the effects of the environment. Lauren Schmitz and Dalton Conley, "Modeling Gene-Environment Interactions With Quasi-Natural Experiments," *Journal of Personality* 85, no. 1 (2017): 10–21, https://doi.org/10.1111/jopy.12227.

38. Anne Case and Angus Deaton, *Deaths of Despair and the Future of Capitalism* (Princeton, NJ: Princeton University Press, 2020), https://press.princeton.edu/books/hardcover/9780691190785/deaths-of-despair-and-the-future-of-capitalism.

39. Case and Deaton.

40. Peter Singer, *A Darwinian Left: Politics, Evolution, and Cooperation* (New Haven, CT: Yale University Press, 2000).

Chapter 9: Using Nature to Understand Nurture

1. Erik Parens, "The Inflated Promise of Genomic Medicine," Scientific American Blog Network, June 1, 2020, https://blogs.scientificamerican.com/observations/the-inflated-promise-of-genomic-medicine/.

2. "Why We Shouldn't Embrace the Genetics of Education," *Just Visiting* (blog), Inside Higher Ed, July 26, 2018, https://www.insidehighered.com/blogs/just-visiting/why-we-shouldnt-embrace-genetics-education.

3. Ruha Benjamin, *Race After Technology: Abolitionist Tools for the New Jim Code* (Cambridge, UK, and Medford, MA: Polity Press, 2019).

4. "WWC | Find What Works!," accessed November 11, 2019, https://ies.ed.gov/ncee/wwc/.

5. "Randomized Controlled Trials Commissioned by the Institute of Education Sciences Since 2002: How Many Found Positive versus Weak or No Effects," Coalition for Evidence-Based Policy, July 2013, http://coalition4evidence.org/wp-content/uploads/2013/06/IES-Commissioned-RCTs-positive-vs-weak-or-null-findings-7-2013.pdf.

6. Hugues Lortie-Forgues and Matthew Inglis, "Rigorous Large-Scale Educational RCTs Are Often Uninformative: Should We Be Concerned?" *Educational Researcher* 48, no. 3 (April 1, 2019): 158–66, https://doi.org/10.3102/0013189X19832850.

7. "Statement of Jon Baron, Vice-President of Evidence-Based Policy, Laura and John Arnold Foundation," House Committee on Agriculture, Subcommittee on Nutrition, July 15, 2015.

8. David S. Yeager et al., "Where and For Whom Can a Brief, Scalable Mindset Intervention Improve Adolescents' Educational Trajectories?," preprint, 2018, accessed November 11, 2019, https://docplayer.net/102132264-Where-and-for-whom-can-a-brief-scalable-mindset-intervention-improve-adolescents-educational-trajectories.html.

9. Laurence Steinberg, "How to Improve the Health of American Adolescents," *Perspectives on Psychological Science* 10, no. 6 (November 1, 2015): 711–15, https://doi.org/10.1177/1745691615598510.

10. Sanjay Srivastava, "Making Progress in the Hardest Science," *The Hardest Science* (blog), March 14, 2009, https://thehardestscience.com/2009/03/14/making-progress-in-the-hardest-science/.

11. "A Different Agenda," *Nature* 487, no. 7407 (July 2012): 271, https://doi.org/10.1038/487271a.

12. Kathryn Paige Harden, "Why Progressives Should Embrace the Genetics of Education," *The New York Times*, July 24, 2018, https://www.nytimes.com/2018/07/24/opinion/dna-nature-genetics-education.html.

13. Benjamin, *Race After Technology*.

14. "Texas Education Code § 28.004," FindLaw, accessed November 11, 2019, https://codes.findlaw.com/tx/education-code/educ-sect-28-004.html.

15. K. Paige Harden, "Genetic Influences on Adolescent Sexual Behavior: Why Genes Matter for Environmentally Oriented Researchers," *Psychological Bulletin* 140, no. 2 (2014): 434–65, https://doi.org/10.1037/a0033564.

16. Felix R. Day et al., "Physical and Neurobehavioral Determinants of Reproductive Onset and Success," *Nature Genetics* 48, no. 6 (June 2016): 617–23, https://doi.org/10.1038/ng.3551.

17. Kathrin F. Stanger-Hall and David W. Hall, "Abstinence-Only Education and Teen Pregnancy Rates: Why We Need Comprehensive Sex Education in the U.S.," *PLoS ONE* 6, no. 10 (October 14, 2011): e24658, https://doi.org/10.1371/journal.pone.0024658.

18. K. Paige Harden et al., "Rethinking Timing of First Sex and Delinquency," *Journal of Youth and Adolescence* 37, no. 4 (April 2008): 373–85, https://doi.org/10.1007/s10964-007-9228-9.

19. Harden, "Genetic Influences on Adolescent Sexual Behavior."

20. Betty Hart and Todd R. Risley, *Meaningful Differences in the Everyday Experience of Young American Children* (Baltimore: Paul H. Brookes Publishing Co., 1995).

21. Clinton Foundation, "Too Small to Fail: Preparing America's Children for Success in the 21st Century," n.d., https://www.clintonfoundation.org/files/2s2f_framingreport_v2r3.pdf.

22. "Empowering Our Children by Bridging the Word Gap," whitehouse.gov, June 25, 2014, https://obamawhitehouse.archives.gov/blog/2014/06/25/empowering-our-children-bridging-word-gap.

23. "About Providence Talks," accessed November 11, 2019, http://www.providencetalks.org/.

24. Douglas E. Sperry, Linda L. Sperry, and Peggy J. Miller, "Reexamining the Verbal Environments of Children From Different Socioeconomic Backgrounds," *Child Development* 90, no. 4 (July/August 2019): 1303–18, https://doi.org/10.1111/cdev.13072.

25. Daniel W. Belsky et al., "The Genetics of Success: How Single-Nucleotide Polymorphisms Associated with Educational Attainment Relate to Life Course Development," *Psychological Science* 27, no. 7 (July 1, 2016): 957–72.

26. Jeremy Freese, "Genetics and the Social Science Explanation of Individual Outcomes," *American Journal of Sociology* 114, suppl. S1 (2008): S1–35, https://doi.org/10.1086/592208.

27. Joseph P. Simmons, Leif D. Nelson, and Uri Simonsohn, "False-Positive Citations," *Perspectives on Psychological Science* 13, no. 2 (March 1, 2018): 255–59, https://doi.org/10.1177/1745691617698146.

28. Freese, "Genetics and the Social Science Explanation of Individual Outcomes."

29. Sam Harris, *Making Sense Podcast* #73, "Forbidden Knowledge," April 22, 2017, https://samharris.org/podcasts/forbidden-knowledge/.

30. "FAQs," Social Science Genetic Association Consortium, accessed March 5, 2019, https://www.thessgac.org/faqs.

31. Sam Trejo and Benjamin W. Domingue, "Genetic Nature or Genetic Nurture? Quantifying Bias in Analyses Using Polygenic Scores," *bioRxiv*, July 31, 2019, 524850, https://doi.org/10.1101/524850.

32. "Dalton Conley," accessed November 11, 2019, https://scholar.princeton.edu/dconley/home.

33. Daniel W. Belsky et al., "Genetic Analysis of Social-Class Mobility in Five Longitudinal Studies," *Proceedings of the National Academy of Sciences* 115, no. 31 (July 31, 2018): E7275–84, https://doi.org/10.1073/pnas.1801238115.

34. Nicholas W. Papageorge and Kevin Thom, "Genes, Education, and Labor Market Outcomes: Evidence from the Health and Retirement Study," NBER Working Paper 25114 (National Bureau of Economic Research, September 2018), https://doi.org/10.3386/w25114.

35. "What Role Should Genetics Research Play in Education?," Stanford Graduate School of Education News, February 20, 2019, https://ed.stanford.edu/news/what-role-should-genetics-research-play-education?print=all.

36. Philipp D. Koellinger and K. Paige Harden, "Using Nature to Understand Nurture," *Science* 359, no. 6374 (January 26, 2018): 386–87, https://doi.org/10.1126/science.aar6429.

37. Augustine Kong et al., "The Nature of Nurture: Effects of Parental Genotypes," *Science* 359, no. 6374 (January 26, 2018): 424–28, https://doi.org/10.1126/science.aan6877.

38. Alicia R. Martin et al., "Clinical Use of Current Polygenic Risk Scores May Exacerbate Health Disparities," *Nature Genetics* 51, no. 4 (April 2019): 584–91, https://doi.org/10.1038/s41588-019-

Chapter 10: Personal Responsibility

1. "Unedited: Amos Wells' Jailhouse Interview," NBC 5 Dallas-Fort Worth, July 3, 2013, https://www.nbcdfw.com/news/local/Unedited-Amos-Wells-Jailhouse-Interview_Dallas-Fort-Worth-214139161.html.

2. Robbie Gonzalez, "How Criminal Courts Are Putting Brains—Not People—on Trial," *Wired*, December 4, 2017, https://www.wired.com/story/how-criminal-courts-are-putting-brains-not-people-on-trial/.

3. Sally McSwiggan, Bernice Elger, and Paul S. Appelbaum, "The Forensic Use of Behavioral Genetics in Criminal Proceedings: Case of the MAOA-L Genotype," *International Journal of Law and Psychiatry* 50 (January–February 2017): 17–23, https://doi.org/10.1016/j.ijlp.2016.09.005.

4. Lisa G. Aspinwall, Teneille R. Brown, and James Tabery, "The Double-Edged Sword: Does Biomechanism Increase or Decrease Judges' Sentencing of Psychopaths?," *Science* 337, no. 6096 (August 17, 2012): 846–49.

5. Nicholas Scurich and Paul Appelbaum, "The Blunt-Edged Sword: Genetic Explanations of Misbehavior Neither Mitigate nor Aggravate Punishment," *Journal of Law and the Biosciences* 3, no. 1 (April 2016): 140–57, https://doi.org/10.1093/jlb/lsv053.

6. Erlend P. Kvaale, William H. Gottdiener, and Nick Haslam, "Biogenetic Explanations and Stigma: A Meta-Analytic Review of Associations among Laypeople," *Social Science & Medicine* 96 (November 2013): 95–103, https://doi.org/10.1016/j.socscimed.2013.07.017.

7. Jeremiah Garretson and Elizabeth Suhay, "Scientific Communication about Biological Influences on Homosexuality and the Politics of Gay Rights," *Political Research Quarterly* 69, no. 1 (March 1, 2016): 17–29, https://doi.org/10.1177/1065912915620050.

8. Fact Sheet Library, NAMI: National Alliance on Mental Illness, accessed November 6, 2019, https://www.nami.org/learn-more/fact-sheet-library.

9. Essi Viding et al., "Evidence for Substantial Genetic Risk for Psychopathy in 7-Year-Olds," *Journal of Child Psychology and Psychiatry* 46, no. 6 (June 2005): 592–97, https://doi.org/10.1111/j.1469-7610.2004.00393.x.

10. American Psychiatric Association, *Diagnostic and Statistical Manual of Mental Disorders,* 4th ed. (Washington, DC: American Psychiatric Association, 2000).

11. Matthew S. Lebowitz, Kathryn Tabb, and Paul S. Appelbaum, "Asymmetrical Genetic Attributions for Prosocial versus Antisocial Behaviour," *Nature Human Behaviour* 3, no. 9 (September 2019): 940–49, https://doi.org/10.1038/s41562-019-0651-1.

12. Lebowitz, Tabb, and Appelbaum. They write, "Our findings add to the substantial body of existing evidence suggesting that factors beyond the inherent quality of biological explanations for behaviour can influence people's likelihood of endorsing them." If "people see genetic explanations as deflecting moral responsibility for behaviour," they reject them "out of a desire to maintain the ability to assign blame."

13. Emily A. Willoughby et al., "Free Will, Determinism, and Intuitive Judgments About the Heritability of Behavior," *Behavior Genetics* 49, no. 2 (March 2019): 136–53, https://doi.org/10.1007/s10519-018-9931-1.

14. Dawkins puts this point well: "Whatever view one takes on the question of determinism, the insertion of the word 'genetic' is not going to make any difference. If you are a full-blooded determinist you will believe that all of your actions are predetermined by physical causes in the past, and you may or may not believe that you therefore cannot be held responsible for your sexual infidelities. But, be that as it may, what difference can it possibly make whether some of the physical causes are *genetic*? Why are genetic determinants thought to be any more ineluctable, or blame-absolving, than environmental ones?" Richard Dawkins, *The Extended Phenotype: The Long Reach of the Gene*, rev. ed. (Oxford and New York: Oxford University Press, 1999).

15. The existence of genomic differences between monozygotic twins means that twin estimates of heritability might be systematically *under*estimated, as phenotypic differences between monozygotic twins caused by genetic differences between them would be misattributed to environmental variation. Hakon Jonsson et al., "Differences between Germline Genomes of Monozygotic Twins," *Nature Genetics* 53, no. 1 (January 2021): 27–34, https://doi.org/10.1038/s41588-020-00755-1.

16. Eric Turkheimer, "Genetics and Human Agency: Comment on Dar-Nimrod and Heine," *Psychological Bulletin* 137, no. 5 (2011): 825–28, https://doi.org/10.1037/a0024306.

17. Daniel C. Dennett, *Elbow Room: The Varieties of Free Will Worth Wanting*, new ed. (Cambridge, MA: MIT Press, 2015).

18. More precisely, *e2* might be thought of as an upper bound of the extent to which people have agency. What the neuroscientist Kevin Mitchell calls "developmental variation," i.e., inherent randomness in processes of phenotypic development, will also pull twins away from one another, without either one of them exerting anything we would typically recognize as agency. Kevin J. Mitchell, *Innate: How the Wiring of Our Brains Shapes Who We Are* (Princeton, NJ: Princeton University Press, 2018).

19. T. J. Bouchard and M. McGue, "Familial Studies of Intelligence: A Review," *Science* 212, no. 4498 (May 29, 1981): 1055–59, https://doi.org/10.1126/science.7195071.

20. Laura E. Engelhardt et al., "Strong Genetic Overlap between Executive Functions and Intelligence," *Journal of Experimental Psychology: General* 145, no. 9 (September 2016): 1141–59, https://doi.org/10.1037/xge0000195.

21. Laura E. Engelhardt et al., "Accounting for the Shared Environment in Cognitive Abilities and Academic Achievement with Measured Socioecological Contexts," *Developmental Science* 22, no. 1 (January 2019): e12699, https://doi.org/10.1111/desc.12699.

22. Kaili Rimfeld et al., "The Stability of Educational Achievement across School Years Is Largely Explained by Genetic Factors," *Npj Science of Learning* 3 (September 4, 2018): 16, https://doi.org/10.1038/s41539-018-0030-0.

23. Amelia R. Branigan, Kenneth J. McCallum, and Jeremy Freese, "Variation in the Heritability of Educational Attainment: An International Meta-Analysis," *Social Forces* 92, no. 1 (September 2013): 109–40.

24. Daniel J. Benjamin et al., "The Promises and Pitfalls of Genoeconomics," *Annual Review of Economics* 4 (September 2012): 627–62, https://doi.org/10.1146/annurev-economics-080511-110939.

25. Dena M. Gromet, Kimberly A. Hartson, and David K. Sherman, "The Politics of Luck: Political Ideology and the Perceived Relationship between Luck and Success," *Journal of Experimental Social Psychology* 59 (July 2015): 40–46, https://doi.org/10.1016/j.jesp.2015.03.002.

26. "Princeton University's 2012 Baccalaureate Remarks," Princeton University, June 3, 2012, https://www.princeton.edu/news/2012/06/03/princeton-universitys-2012-baccalaureate-remarks.

27. Jonathan Rothwell, "Experiment Shows Conservatives More Willing to Share Wealth Than They Say," *The New York Times*, February 13, 2020, https://www.nytimes.com/2020/02/13/upshot/trump-supporters-experiment-inequality.html.

28. Heather MacDonald, "Who 'Deserves' to Go to Harvard?," *Wall Street Journal*, June 13, 2019, https://www.wsj.com/articles/who-deserves-to-go-to-harvard-11560464201.

29. Quoted in James Pethokoukis, "You Didn't Build That: Obama and Elizabeth Warren Argue against Any Limiting Principle to Big Government," blog post, *AEIdeas*, American Enterprise Institute, July 19, 2012, https://www.aei.org/pethokoukis/you-didnt-build-that-obama-and-elizabeth-warren-argue-against-any-limiting-principle-to-big-government/.

30. Stephen P. Schneider, Kevin B. Smith, and John R. Hibbing, "Genetic Attributions: Sign of Intolerance or Acceptance?," *The Journal of Politics* 80, no. 3 (July 2018): 1023–27, https://doi.org/10.1086/696860.

31. Rothwell, "Experiment Shows Conservatives More Willing to Share Wealth Than They Say."

32. Ingvild Almås et al., "Fairness and the Development of Inequality Acceptance," *Science* 328, no. 5982 (May 28, 2010): 1176–78, https://doi.org/10.1126/science.1187300; Alexander W. Cappelen, Erik Ø. Sørensen, and Bertil Tungodden, "Responsibility for What? Fairness and Individual Responsibility," *European Economic Review* 54, no. 3 (April 2010): 429–41, https://doi.org/10.1016/j.euroecorev.2009.08.005; Alexander W. Cappelen et al., "Just Luck: An Experimental Study of Risk-Taking and Fairness," *The American Economic Review* 103, no. 4 (2013): 1398–1413.

33. Ingvild Almås, Alexander W. Cappelen, and Bertil Tungodden, "Cut-throat Capitalism versus Cuddly Socialism: Are Americans More Meritocratic and Efficiency-Seeking than Scandinavians?," *Journal of Political Economy* 128, no. 5 (May 2020): 1753–88, https://doi.org/10.1086/705551.

34. Michael Young, "Down with Meritocracy," *The Guardian*, June 28, 2001, https://www.theguardian.com/politics/2001/jun/29/comment.

Chapter 11: Difference without Hierarchy

1. "Homelessness and Mental Illness: A Challenge to Our Society," Brain & Behavior Research Foundation, November 19, 2018, https://www.bbrfoundation.org/blog/homelessness-and-mental-illness-challenge-our-society.

2. Erik Parens, "Genetic Differences and Human Identities. On Why Talking about Behavioral Genetics Is Important and Difficult," *The Hastings Center Report* special supplement 34, no. 1 (January-February 2004): S14–36, https://www.thehastingscenter.org/wp-content/uploads/genetic_differences_and_human_identities.pdf.

3. Elizabeth S. Anderson, "What Is the Point of Equality?," *Ethics* 109, no. 2 (January 1999): 287–337, https://doi.org/10.1086/233897.

4. Audre Lorde, "Reflections," *Feminist Review* 45 (Autumn 1993): 4–8.

5. Daniel J. Kevles, *In the Name of Eugenics: Genetics and the Uses of Human Heredity* (New York: Alfred A. Knopf, 1985; reprint, Cambridge, MA: Harvard University Press, 1998).

6. Henry Herbert Goddard, *Feeble-Mindedness: Its Causes and Consequences* (New York: Macmillan, 1914).

7. Nathaniel Comfort, "How Science Has Shifted Our Sense of Identity," *Nature* 574, no. 7777 (October 2019): 167–70, https://doi.org/10.1038/d41586-019-03014-4.

8. "Excuse Me, Mr Coates, Ctd," *The Dish*, December 13, 2014, http://dish.andrewsullivan.com/2014/12/23/excuse-me-mr-coates-ctd/.

9. Ibram X. Kendi, *How to Be an Antiracist* (New York: One World, 2019).

10. Douglas Almond, Kenneth Y. Chay, and Michael Greenstone, "Civil Rights, the War on Poverty, and Black-White Convergence in Infant Mortality in the Rural South and Mississippi," MIT Department of Economics Working Paper no. 07-04, SSRN (Rochester, NY: Social Science Research Network, February 7, 2007), https://papers.ssrn.com/abstract=961021.

11. "Flint, Michigan, Decision to Break Away from Detroit for Water Riles Residents," CBS News," March 4, 2015, https://www.cbsnews.com/news/flint-michigan-break-away-detroit-water-riles-residents/.

12. Mona Hanna-Attisha et al., "Elevated Blood Lead Levels in Children Associated With the Flint Drinking Water Crisis: A Spatial Analysis of Risk and Public Health Response," *American Journal of Public Health* 106, no. 2 (February 2016): 283–90, https://doi.org/10.2105/AJPH.2015.303003.

13. Michigan Civil Rights Commission, *The Flint Water Crisis: Systemic Racism through the Lens of Flint*, February 17, 2017, https://www.michigan.gov/documents/mdcr/VFlintCrisisRep-F-Edited3-13-17_554317_7.pdf.

14. Harriet A. Washington, *A Terrible Thing to Waste: Environmental Racism and Its Assault on the American Mind* (New York: Little, Brown Spark, 2019).

15. Washington.

16. A. Alexander Beaujean et al., "Validation of the Frey and Detterman (2004) IQ Prediction Equations Using the Reynolds Intellectual Assessment Scales," *Personality and Individual Differences* 41, no. 2 (July 2006): 353–57, https://doi.org /10.1016/j.paid.2006.01.014.

17. Catherine M. Calvin et al., "Intelligence in Youth and All-Cause-Mortality: Systematic Review with Meta-Analysis," *International Journal of Epidemiology* 40, no. 3 (June 1, 2011): 626–44, https://doi.org/10.1093/ije/dyq190.

18. Meredith C. Frey and Douglas K. Detterman, "Scholastic Assessment or g? The Relationship between the Scholastic Assessment Test and General Cognitive Ability," *Psychological Science* 15, no. 6 (June 1, 2004): 373–78, https://doi.org/10 .1111/j.0956-7976.2004.00687.x.

19. Christopher M. Berry and Paul R. Sackett, "Individual Differences in Course Choice Result in Underestimation of the Validity of College Admissions Systems," *Psychological Science* 20, no. 7 (July 1, 2009): 822–30, https://doi.org/10.1111/j .1467-9280.2009.02368.x.

20. David Lubinski and Camilla Persson Benbow, "Study of Mathematically Precocious Youth After 35 Years: Uncovering Antecedents for the Development of Math-Science Expertise," *Perspectives on Psychological Science* 1, no. 4 (December 1, 2006): 316–45, https://doi.org/10.1111/j.1745-6916.2006.00019.x.

21. Ann Oakley, "Gender, Methodology and People's Ways of Knowing: Some Problems with Feminism and the Paradigm Debate in Social Science," *Sociology* 32, no. 4 (November 1, 1998): 707–31, https://doi.org/10.1177/0038038598032004005.

22. Kevin Cokley and Germine H. Awad, "In Defense of Quantitative Methods: Using the 'Master's Tools' to Promote Social Justice," *Journal for Social Action in Counseling & Psychology* 5, no. 2 (Summer 2013): 26–41.

23. Carol A. Padden and Tom L. Humphries, *Deaf in America: Voices from a Culture* (Cambridge, MA: Harvard University Press, 1988).

24. Abraham M. Sheffield and Richard J. H. Smith, "The Epidemiology of Deafness," *Cold Spring Harbor Perspectives in Medicine* 9, no. 9 (September 3, 2019): a033258, https://doi.org/10.1101/cshperspect.a033258.

25. Walter E. Nance, "The Genetics of Deafness," *Mental Retardation and Developmental Disabilities Research Reviews* 9, no. 2 (2003): 109–19, https://doi.org/10 .1002/mrdd.10067.

26. M. Spriggs, "Lesbian Couple Create a Child Who Is Deaf like Them," *Journal of Medical Ethics* 28, no. 5 (October 2002): 283, https://doi.org/10.1136/jme.28.5.283.

27. Isabel Karpin, "Choosing Disability: Preimplantation Genetic Diagnosis and Negative Enhancement," *Journal of Law and Medicine* 15, no. 1 (August 2007): 89–103.

28. Steven D. Emery, Anna Middleton, and Graham H. Turner, "Whose Deaf Genes Are They Anyway?: The Deaf Community's Challenge to Legislation on Embryo Selection," *Sign Language Studies* 10, no. 2 (2010): 155–69.

29. "This Couple Want a Deaf Child. Should We Try to Stop Them?" *The Guardian*, March 9, 2008, https://www.theguardian.com/science/2008/mar/09 /genetics.medicalresearch.

30. H. Dominic W. Stiles and Mina Krishnan, "What Happened to Deaf People during the Holocaust?," UCL Ear Institute & Action on Hearing Loss Libraries,

University College London, November 16, 2012, https://blogs.ucl.ac.uk/library-rnid/2012/11/16/what-happened-to-deaf-people-during-the-holocaust/.

31. Paul Steven Miller and Rebecca Leah Levine, "Avoiding Genetic Genocide: Understanding Good Intentions and Eugenics in the Complex Dialogue between the Medical and Disability Communities," *Genetics in Medicine* 15, no. 2 (February 2013): 95–102, https://doi.org/10.1038/gim.2012.102; Emery, Middleton, and Turner, "Whose Deaf Genes Are They Anyway?"

32. Anderson, "What Is the Point of Equality?"

33. John Rawls, *A Theory of Justice*, rev. ed. (Cambridge, MA: Harvard University Press, 1999).

34. David Kushner, "Serving on the Spectrum: The Israeli Army's Roim Rachok Program Is Bigger Than the Military," *Esquire*, April 2, 2019, https://www.esquire.com/news-politics/a26454556/roim-rachok-israeli-army-autism-program/.

35. Robert D. Austin and Gary P. Pisano, "Neurodiversity as a Competitive Advantage," *Harvard Business Review*, May-June 2017, https://hbr.org/2017/05/neurodiversity-as-a-competitive-advantage.

36. Susan Dominus, "Open Office," *The New York Times Magazine*, February 21, 2019, https://www.nytimes.com/interactive/2019/02/21/magazine/autism-office-design.html, https://www.nytimes.com/interactive/2019/02/21/magazine/autism-office-design.html.

37. John Elder Robison, "What Is Neurodiversity?," *My Life with Asperger's* (blog), *Psychology Today*, October 7, 2013, http://www.psychologytoday.com/blog/my-life-aspergers/201310/what-is-neurodiversity.

Chapter 12: Anti-Eugenic Science and Policy

1. Elizabeth S. Anderson, "What Is the Point of Equality?," *Ethics* 109, no. 2 (January 1999): 287–337, https://doi.org/10.1086/233897.

2. Ibram X. Kendi, *How to Be an Antiracist* (New York: One World, 2019).

3. Ruha Benjamin, ed., *Captivating Technology: Race, Carceral Technoscience, and Liberatory Imagination in Everyday Life* (Durham, NC: Duke University Press, 2019).

4. Mark A. Rothstein, "Legal Conceptions of Equality in the Genomic Age," *Law & Inequality* 25, no. 2 (2007): 429–63.

5. Eric Turkheimer, "Three Laws of Behavior Genetics and What They Mean," *Current Directions in Psychological Science* 9, no. 5 (October 1, 2000), 160–64, https://journals.sagepub.com/doi/10.1111/1467-8721.00084.

6. Theodosius Dobzhansky, "Genetics and Equality: Equality of Opportunity Makes the Genetic Diversity among Men Meaningful," *Science* 137, no. 3524 (July 13, 1962): 112–15, https://doi.org/10.1126/science.137.3524.112.

7. Antonio Regalado, "DNA Tests For IQ Are Coming, But It Might Not Be Smart to Take One," *MIT Technology Review*, April 2, 2018, https://getpocket.com/explore/item/dna-tests-for-iq-are-coming-but-it-might-not-be-smart-to-take-one.

8. Robert Plomin, *Blueprint: How DNA Makes Us Who We Are* (Cambridge, MA: MIT Press, 2018).

9. Tim T. Morris, Neil M. Davies, and George Davey Smith, "Can Education Be Personalised Using Pupils' Genetic Data?" *bioRxiv*, December 11, 2019, 645218, https://doi.org/10.1101/645218.

10. Safiya Umoja Noble, *Algorithms of Oppression: How Search Engines Reinforce Racism* (New York: NYU Press, 2018); Cathy O'Neil, *Weapons of Math Destruction: How Big Data Increases Inequality and Threatens Democracy*, repr. ed. (New York: Broadway Books, 2017).

11. Julia Angwin et al., "Machine Bias," ProPublica, May 23, 2016, https://www .propublica.org/article/machine-bias-risk-assessments-in-criminal-sentencing.

12. Benjamin, *Captivating Technology*.

13. O'Neil, *Weapons of Math Destruction*; Noble, *Algorithms of Oppression*.

14. Sean F. Reardon, "School District Socioeconomic Status, Race, and Academic Achievement," Stanford Center for Education Policy Analysis (CEPA), April 2016, https://cepa.stanford.edu/content/school-district-socioeconomic -status-race-and-academic-achievement.

15. Stephen W. Raudenbush and J. Douglas Willms, "The Estimation of School Effects," *Journal of Educational and Behavioral Statistics* 20, no. 4 (Winter 1995): 307–35, https://doi.org/10.3102/10769986020004307.

16. K. Paige Harden et al., "Genetic Associations with Mathematics Tracking and Persistence in Secondary School," *Npj Science of Learning* 5 (February 5, 2020): 1, https://doi.org/10.1038/s41539-020-0060-2.

17. Robert Moses, "Math As a Civil Rights Issue: Working the Demand Side," Harvard Gazette, May 17, 2001, https://news.harvard.edu/gazette/story/2001/05 /math-as-a-civil-rights-issue/.

18. Lorie Konish, "This Is the Real Reason Most Americans File for Bankruptcy," CNBC, February 11, 2019, https://www.cnbc.com/2019/02/11/this-is-the -real-reason-most-americans-file-for-bankruptcy.html.

19. "Genetic Discrimination," National Human Genome Research Institute, accessed March 10, 2020, https://www.genome.gov/about-genomics/policy -issues/Genetic-Discrimination.

20. Mark A. Rothstein, "GINA at Ten and the Future of Genetic Nondiscrimination Law," *The Hastings Center Report* 48, no. 3 (May/June 2018): 5–7, https://doi .org/10.1002/hast.847.

21. Jessica L. Roberts, "The Genetic Information Nondiscrimination Act as an Antidiscrimination Law," *Notre Dame Law Review* 86, no. 2 (2013): 597–648, http://ndlawreview.org/wp-content/uploads/2013/06/Roberts.pdf.

22. Rothstein, "Legal Conceptions of Equality in the Genomic Age."

23. Mark A. Rothstein, "Why Treating Genetic Information Separately Is a Bad Idea," *Texas Review of Law & Politics* 4, no. 1 (Fall 1999): 33–37.

24. Mark A. Rothstein, "Genetic Privacy and Confidentiality: Why They Are So Hard to Protect," *Journal of Law, Medicine and Ethics* 26, no. 3 (Fall 1998): 198–204, https://papers.ssrn.com/abstract=1551287.

25. Roberts, "The Genetic Information Nondiscrimination Act as an Antidiscrimination Law."

26. John Rawls, *A Theory of Justice*, rev. ed. (Cambridge, MA: Harvard University Press, 1999).

27. Robert H. Frank, *Success and Luck: Good Fortune and the Myth of Meritocracy* (Princeton, NJ: Princeton University Press, 2016).

28. David Roberts, "The Radical Moral Implications of Luck in Human Life," *Vox*, August 21, 2018, https://www.vox.com/science-and-health/2018/8/21/17687402/kylie-jenner-luck-human-life-moral-privilege.

29. Amartya Sen, "Merit and Justice," in *Meritocracy and Economic Inequality*, ed. Kenneth J. Arrow, Samuel Bowles, and Steven Durlauf (Princeton, NJ: Princeton University Press, 2000).

30. Madeleine L'Engle, *A Wrinkle in Time*, reprint ed. (New York: Square Fish, 2007).

31. Rawls, *A Theory of Justice*.

32. Angus Deaton, *The Great Escape: Health, Wealth, and the Origins of Inequality* (Princeton, NJ: Princeton University Press, 2013).

33. François Bourguignon and Christian Morrisson, "Inequality Among World Citizens: 1820–1992," *American Economic Review* 92, no. 4 (September 2002): 727–44, https://doi.org/10.1257/00028280260344443.

34. Max Roser, Hannah Ritchie, and Bernadeta Dadonaite, "Child and Infant Mortality," *Our World in Data*, May 10, 2013, https://ourworldindata.org/child-mortality; "Sweden: Child Mortality Rate 1800-2020," Statista, accessed February 9, 2021, https://www.statista.com/statistics/1041819/sweden-all-time-child-mortality-rate/.

35. Daron Acemoglu, "Technical Change, Inequality, and the Labor Market," *Journal of Economic Literature* 40, no. 1 (March 2002): 7–72.

36. Heather MacDonald, "Who 'Deserves' to Go to Harvard?," *Wall Street Journal*, June 13, 2019, https://www.wsj.com/articles/who-deserves-to-go-to-harvard-11560464201.

37. Anne Case and Angus Deaton, *Deaths of Despair and the Future of Capitalism* (Princeton, NJ: Princeton University Press, 2020), https://press.princeton.edu/books/hardcover/9780691190785/deaths-of-despair-and-the-future-of-capitalism.

INDEX

A NOTE ON THE TYPE

This book has been composed in Adobe Text and Gotham.
Adobe Text, designed by Robert Slimbach for Adobe,
bridges the gap between fifteenth- and sixteenth-century
calligraphic and eighteenth-century Modern styles.
Gotham, inspired by New York street signs, was designed
by Tobias Frere-Jones for Hoefler & Co.